Pictorial Mathematics

An Engaging Visual Approach To The Teaching And Learning Of Mathematics

Guillermo Mendieta Jr.

Meaningful Learning
P.O. Box 448. - Etiwanda, Ca 91739
PictorialMath.com

Pictorial Mathematics

A Visual Approach
To The Teaching And Learning Of Mathematics

By Guillermo Mendieta

Published by:
Meaningful Learning
PO Box 448
Etiwanda, CA 91739 U.S.A.
Orders and professional development information at: PictorialMath.com

All rights reserved. No part of this book may be reproduced or transmitted in any form or by any means, electronic or mechanical, including photocopying, recording or by any information storage and retrieval system, without written permission from the author, except for for the inclusion of brief quotations for a review. Individual teachers who purchase the book may make copies for his or her students.

Copyright © 2005 – by Guillermo Mendieta

ISBN, 0-9773212-8-2

First printing 2006

Printed in the United States of America

Editor: Abigail Mieko Vargus

Cover design/Art direction: Jana Rade

Illustrator: Lilian Barac

Library of Congress Control Number: 2005937547

*"Learning
is not a spectator sport."*

- D. Blocher

Table of Contents

Who should get a copy of this resource book	…………………..	10
About the Author	…………………..	11
Preface	…………………..	13
Acknowledgements	…………………..	18
Introduction	…………………..	21
Methodology	…………………..	31
Number and Operations	…………………..	35
What Will Students Be Conceptualizing?	…………………..	36
Teacher Notes	…………………..	38
Pictorial Place Value - Whole Numbers	…………………..	41
Adding Whole Numbers	…………………..	45
Teacher Notes	…………………..	49
Subtraction - Whole Numbers	…………………..	50
Thinking Through Basic Skills	…………………..	53
Teacher Notes	…………………..	56
Adding Positive and Negative Integers	…………………..	58
Multiplication	…………………..	61
Teacher Notes	…………………..	62
Developing Multiplication Understanding	…………………..	64
Multiplication Models	…………………..	69
Recognizing Multiplication	…………………..	70
Linking the Standard Multiplication to a Pictorial Representation	…………………..	73
Introducing 2-Digit Pictorial Multiplications	…………………..	74
Base-10 Double Digit Multiplication	…………………..	75
Box Multiplication	…………………..	78

Understanding the Process of Multiplying
and Dividing Operators 81
Factor Trees and Factor Ladders 84

Decimals 87
What Will Students Be Conceptualizing? 88
Teacher Notes 90
Pictorial Decimal Representations 92
Pictorial Place Value - Decimals 96
Pictorial Decimal/Fractional Representations 100
Decimal Shadings 103
Percent Shadings 105
Estimating Decimals 108
Estimating Percents 110
Comparing Decimals and Fractions 113
10 by 10 Shadings 116
Pictorial/Symbolic Decimal Addition 119
Pictorial/Symbolic Decimal Multiplication 122
Pictorial Decimal Multiplication/Factoring 125
Algebraic Decimal Representations 130

Fractions 133
What Will Students Be Conceptualizing? 134
Pictorial Templates Most Useful for Teaching
And Learning Fractions 136
Teacher Notes 137
Developing Fraction Understanding 141
Multiple Representations of Simple Fractions 144
Fraction Match 147
Missing Numerators and Denominators 149

Developing Fraction Understanding - Changing Units 151
Estimating Fractions 154
From Fractions to Units 156
Fraction Partitioning 159
Fraction Size Sense (overheads) 161
Adding Fractions, Common Denominators 164
Adding Fractions, Uncommon Denominators 167
Subtracting Fractions, Common Denominators 172
Subtracting Fractions, Uncommon Denominators 174
Repeating Groups 178
Multiplication of Mixed Fractions 181
Multiplication of Fractions 184
Teacher Notes 187
Division of Fractions 190
Developing Percent Understanding - Changing Units 193
Relation-Strips 196
From Percents to Units 199
Developing Fraction / Percent Connections 202
Fraction Card Games 203

Ratios 217
Developing Ratio Understanding 218
Shading Ratios 221
Circle Ratios 223
Developing Ratio Sense 225
Grid Ratios 227
Ratios vs. Fractions 230
Pictorial Transformations 233

Algebra 235
- What Will Students Be Conceptualizing? 236
- Expression Representation 238
- Representing Unknowns 242
- Pictorial Algebraic Representations 243
- Adding Expressions 247
- Subtracting Expressions 250
- Algebraic Number Line 253
- Algebraic Area 256
- Multiplying Expressions 262
- Shading-in Expressions 264
- Factoring 268
- Matching Expressions 276
- Linear Equation Match 278
- Algebraic Linear Measurement 281
- Algebra Card Games 284

Geometry 295
- Constructing Figures 296
- Area - Perimeter Connections 297
- Find the Perimeter 300
- Algebraic Perimeter 301
- Draw the Area 302
- Find the Area 303
- Reflections 305
- Geometrical Compass 307
- Isometric Copies 309
- Geometrical Directions 310
- Geometrical Analogies 311

   Geometrical Errors 313
   True or False Geo 314
   Volume 315
   Squares Within Squares 317
   Fold-it 3-Dimensional Polyhedrons 319

Probability and Statistics 329
   Creating a Fair Spinner 330
   Fair Spinner? 331
   Carnival Spinner 332
   Dart's Chances 333
   Average U.S. Monthly Income by Years of Study 334
   Graphing Transformations 335
   How Many Students Brought Their Homework? 336
   How High Does the Ball Bounce? 338
   Birthday Graph 339
   Counting and Eating M&M's 341
   How Large is your Shoe? 342

Appendix 1 - Pictorial Templates 343
   10 by 10 Grids 345
   Graphing Grids 348
   1-Inch Graph Paper 349
   Circle Fractions 350
   24 Hour Clock 353
   Place Value Mats 354
   Base-10 Manipulatives 356
   Geoshapes 357
   Large 6 by 4 Half of Half Grid 358
   4 by 4 Dot Paper 359

Half-Inch Graph Paper	360
Quarter-Inch Square Paper	361
Isometric Paper	362
Number Lines	364
Pictorial Worksheets	365
Grids	366
Number Charts	368
Bar Graph Grids	371
60 Second Clock	372
Geoboard Paper	373
Dot Picture Paper	374
Twelve Circles	375
4 by 3 Rectangular Dot Paper	376
Large 4 by 3 Grid	377
Large 6 by 4 Grid	378
Tables	379
Sample Activities/Tasks Based On Templates	383

Bibliography 393

Workshop and Training Information 397

Quick Order Form 399

Quick Order Form[1]

Make checks payable to **Meaningful Learning**

Order online at www.PictorialMath.com or use this form to order by fax or mail

Mail or Fax to:
Meaningful Learning
P.O. Box 448
Etiwanda, CA 92336
(909) 730-7312 – Fax (909) 909-854-5858

Order online before February 15th and save even more!

We accept checks, credit cards and purchase orders (purchase orders must be faxed or mailed).

1-4 books:	$34.95 per book
5-15 books:	$31.45 per book (10% discount)
16-30 books:	$29.70 per book (15% discount)
31+ books	Contact us

District orders of 100 books or more qualify for the free professional development offer. Call or email us for further details.

You can order online at PictorialMath.com

Name: _____

Institution: _____

Address: _____

City _____ State _____

Zip code: _____ Telephone: _____

Email: _____ Fax: _____

Number of Books	Price/Book (as detailed above)	Method of Payment	Subtotal
Taxes (add 7.75% if ordering within California - $2.70 for 1 book)			
Shipping* (3.50 for first book, $2.50 for every additional book)			
		Total	

www.PictorialMath.com

About the Author

Guillermo Mendieta published his first mathematics book at the age of eighteen while earning his bachelors of science in pure mathematics at Louisiana State University. He is currently the director of Meaningful Learning, a professional development organization working with districts and schools to improve teaching, learning and instructional leadership in mathematics. During the last fifteen years Mr. Mendieta has been the director of several mathematics initiatives with The Achievement Council in Los Angeles, California. He has taught mathematics at all grades from sixth-grade through college level and has served as a mathematics coach for several schools throughout California.

Guillermo has worked with thousands of teachers and dozens of districts across the country to improve student achievement in mathematics. He was the co-director of California's first statewide conference on mathematics and the Latino student. He is the recipient of the NAACP Los Angeles Teacher of the Year award, as well as one of the graduate student research awards from California State University, Los Angeles.

Mr. Mendieta was a contributing author to the Any-Time Math Series. He has served in numerous boards, including the advisory boards of the California Math Council, the National Science Foundation's Systemic Reform Initiatives, and the Annenberg Foundation's National Equity Advisory Board. Guillermo is a certified AB466 trainer for the Algebra McDougal Littell and the Prentice Hall textbook series. He is a sought-after, dynamic and engaging professional developer and keynote speaker and is a certified group facilitator. Guillermo is currently working on the second volume of pictorial mathematics.

*I dedicate this book
to my mother and father.
They taught me
the value and power
of education.*

*"Do not worry
about your difficulties in mathematics,
I assure you that mine are greater."*

Albert Einstein

Preface

The publication of this book coincides with one of the most challenging and crucial times in educators' efforts to significantly reform mathematics education in the United States. Teachers, the most crucial players in the mathematics reform equation, find themselves in the midst of a seemingly impossible dichotomy: align their teaching practices with the recommendations of the best research on teaching and learning which call for emphasizing conceptual understanding and problem solving, and/or emphasize the automatization of basic skills on which most students are tested, and so many found lacking.

The current pressure for teachers to focus on getting students to memorize facts and procedures is intense. Most states use standardized tests that overwhelmingly favor those students who can quickly recall facts and procedures. Students are promoted or retained based in large part on these test scores. In addition, teachers and principals are evaluated based on their students' test performance, with some teachers and principals receiving bonuses as large as $25,000 for raising test scores. To make it even tougher for teachers, despite an overwhelming recognition that in the United States we teach a mile-long curriculum at an inch of depth, most state standards still dictate teachers to teach such a long list of concepts and skills that most teachers find time for only superficial coverage of most topics. So, what is a teacher to do?

This book was created to provide teachers with a bridge between basic skills, conceptual development and problem-solving skills. The collection of pictorial tasks is intended as a bridge between concrete and more symbolic representations. This resource will support teachers' efforts to teach mathematics meaningfully, with a focus on developing students' conceptual understanding. The author has purposefully devoted over 85% of the book's 400 pages to providing teachers with models, templates and exercises that they can immediately use in their classroom.

The pictorial tasks in this book are not meant to replace the more concrete activities that students need in the early stages of introducing a mathematical concept. Instead, they are an appropriate next level of abstraction after the students have gone through Bruner's (1960) enactive learning stage, or Dienes' (1969) play-and-exploration stage with manipulatives. Fig. 3 shows one example of the type of instructional sequence that is aligned with the theoretical framework that supports the approach envisioned for the tasks in this resource book.

Fig 3.

Sample Instructional Sequence for Number Representation

Stage	What The Teacher Does	What Students Do
Play - Explore	The teacher gives students Base-10 blocks to play and explore.	Students make designs, construct patterns, etc., using the base-10 blocks.
Guiding Questions	The teacher asks some open-ended questions (i.e. what can you tell me about the different blocks?)	Students talk with their partners, record their observations on a graphic organizer, share with their findings with the class, etc.
Field-Specific Questions	The teacher asks more content-specific questions, requiring students to manipulate and construct numbers with the blocks.	After establishing that ▢ = 100 ▯ =10 ▫ =1 the students construct the numbers (132) (213), and (321)

Formative Reflective Questions	The teacher asks reflective questions after several constructions (i.e. what can you tell me about the value of the digit 3 in each of the numbers you constructed? How did it change?)	Students talk with their partners, record their observations on a graphic organizer, share their thinking with the class, etc.
Transformation Tasks	The teacher asks students to share different ways of constructing a given number with the blocks.	Students construct the number 12 with twelve ones or with 1 ten-block and two ones. They talk with their partners, share with the class, etc.
Translation Tasks	The teacher gives students one type of representation for a number (i.e. pictorial, oral, written words, standard numeric, expanded notation, etc.) and they are asked/taught to construct the other types.	Students represent the number 12, as twelve, 10+2, 11+1, ▯ ▫ ▫ 1(10) + 2(1), one dozen, etc.
Summative Reflective Questions	The teacher asks reflective questions to summarize and connect the ideas learned - (i.e. how is each representation connected to other ideas you have learned before?)	Students practice summarizing with their partners, record their observations on a graphic organizer, share their ideas with the class, etc.

The difference between transformations and translations is briefly explained in the introduction. For now, I invite the reader to try to figure-out the differences from the example given.

I began thinking about this book after I *read Problems of Representation in the Teaching and Learning of Mathematics*, edited by Claude Janvier, back in the early 1990s. As I worked with teachers, it became clear that there was a need for a practitioner to articulate the research ideas about the important role that the construction, translation and transformations of representations play in the teaching and learning of mathematics. I wanted to create teacher-friendly materials that put these ideas into practice and gave teachers a way to reconceptualize "drill and practice" in the context of developing conceptual understanding. My approach can be summarized as follows:

- Focus on the most important ideas in second through ninth-grade mathematics, not on the coverage of a long list of prescribed concepts and skills.

- Provide models of the type of questions, language, pictures and solutions that would illustrate an effective translation and transformational problem-solving environment.

- Provide a substantial number of exercises designed to promote higher levels of conceptual development.

- Provide a way for teachers to create their own translation and transformation tasks through the master and pictorial templates.

- Suggest examples of the type of language and the inner thinking that would make the pictorial models most useful.

- Create enough pictorial models across several mathematical concepts so as to promote and facilitate the transfer of the use of pictures from the margins of instruction to a more prominent role in daily instructional practice.

- Organize the exercises in a way that, with the teacher's guidance, helps students internalize the structural features of the pictorial model - that is, helps students become more proficient at abstracting and generalizing.

The systematic and purposeful introduction of translations and transformations of pictorial representations within the main topics of second through ninth-grade serves, as Kaput (1998) puts it, to "algebrafy" the elementary and middle school curriculum. Given the systemic large-scale failure rates in algebra classes that currently exists, and the critical college gatekeeper role that algebra plays in our system, we need to create easily adaptable systemic scaffolds that prepare students to succeed in high school mathematics and beyond. The fact that the exercises and the models presented in this resource "look familiar" to teachers may make them more likely to be incorporated as part of routinely used instructional representations.

I hope educators will find these materials useful in helping students develop stronger conceptual understanding of mathematical concepts. I invite readers to write me at pictorialmath@yahoo.com to share their experiences with the materials and to make suggestions for future volumes or editions.

Acknowledgments

Few things in this world can be done well without substantial help from others. After three years of working on this book, I want to thank those who directly and indirectly contributed to the creation and development of this book.

First and foremost, I want to thank my wife Elsa for her candid feedback, her continuous support, her immense patience, and most of all, the inspiration to do my best. Along with my wife, I want to thank my children, Anthony, David and Carlos. Being patient and kind, my children were willing to work on the problems I created, and their reaction, both positive and negative, helped me create a better book.

I want to express my deepest thanks to eight people for their invaluable contributions to this book. They took time from their extremely busy schedules to read the manuscript at different stages of development and give me numerous suggestions for improvement and much needed words of encouragement. Dr. Robert E. Reys, Curators' professor of mathematics education at the University of Missouri-Columbia; Dr. Dorothy Keane, professor of mathematics education at California State University, Los Angeles; K.C. Cole, Los Angeles Times science writer and author of The Universe and the Teacup: The Mathematics of Truth and Beauty; Richard Lee Colvin, director of the Hechinger Institute on Education and the Media, Teachers College, Columbia University; Gerry Mendieta, my brother and math teacher at Carson high school; Carlos Lemus, executive director of the Coalition for Educational Partnerships; Vance Mills, School-based services manager, California Gear up. A special thanks to the late Dr. James Kaput, chancellor professor of Mathematics at Dartmouth University, whose research inspired many of the ideas in the book and who gracefully agreed to review the book. My deepest sympathy goes out to his family and friends for his untimely passing.

I want to acknowledge Jana Rade for her excellent cover design and art direction. Jana, along with Lilian Barak, the cover illustrator, were able to capture my vision of the book as a bridge between abstract and concrete ways of teaching mathematics. I also want to thank Abigail Mieko Vargus, my editor. Abigail's editing skills helped me improve the quality of the narrative portions of the book

I want to thank the teachers at Florence Elementary and Gage Middle Schools in the Los Angeles Unified School District for testing the materials, giving me feedback and inspiring several of the ideas presented in the book. I also want to thank Dr. Dale Vigil, Pat Forkas Mckenna, and Glynn Thompson, administrators in District 6 in the Los Angeles Unified School District, for their support and for setting such a high standard of educational leadership.

I want to acknowledge four people who greatly influenced my approach to the teaching and learning of mathematics. Dr. Dorothy Keane, my graduate mentor at California State University, Los Angeles, inspired me to relearn mathematics with meaning, purpose and understanding. I want to thank Richard Curci. His boundless enthusiasm and support have made such a difference in my life. From the Mathematics Renaissance, to the Achievement Council, his hands have been all over my career. Phyllis Hart, executive director of The Achievement Council during my thirteen-year tenure with the Council. I want to thank her for the many learning opportunities she made available to me. I want to extend a special thanks to Guadalupe Simpson, my principal at Nimitz Middle School. She gave me the space to spread my wings and grow. Thank you for always standing up for what was best for children, even when it was at a high cost to you.

I want to thank Pam Good, my colleague and friend at The Achievement Council. I am thankful that she was there to support the organization during the tough times. I want to acknowledge Tony Alvarado, the best superintendent in the nation during my thirteen-year tenure at the Achievement Council. From Tony I learned what was possible to be accomplished when instructional leadership was strong, front and center.

Finally, I want to thank my father, Guillermo Mendieta Sr., and my mother, Gerda Mendieta Schiebel. You both lead such extraordinary lives, full of courage, dedication and sacrifice. You gave your three children a thirst for learning and a deep respect for education. I dedicate this book to both of you.

*A student's depth
of conceptual understanding
is directly related
to their ability to translate
and transform concepts
within and across
representational systems.*

- G. Mendieta

A picture is worth a thousand words.

Introduction

This book was created to provide teachers with a bridge between basic skills, conceptual development and problem solving skills. Its purposeful design uses pictures as the bridge between concrete and abstract representations. The author intends for Pictorial Mathematics to be used as a companion to most school mathematics textbooks in grades three through nine.

Mastery of the key mathematic standards in school mathematics requires a fairly deep conceptual understanding of the key mathematical ideas. To gain that understanding, teachers and students can build a pictorial mental framework based on the collection of exercises, problems and pictorial models in this book.

The premise of this book is grounded on the research of the role that multiple representations plays in the development of conceptual understanding. This research suggests that the level or depth of anyone's conceptual understanding of a mathematical concept is strongly linked to three key variables:

- ❑ The number, appropriateness and strength of the connections made between the concept being studied and other ideas held by the learner

- ❑ The learner's ability to work with a wide variety of representations of the concept within different contexts.

❑ The learner's ability to select and use an efficient representation given the problem situation

This resource uses the research on multiple representations to give teachers effective tools to develop students' conceptual understanding of the key mathematical concepts across the strands. Mathematical concepts are inherently symbolic in nature, and as such, they are represented with a variety of symbols, graphs, charts, etc. The type and sequencing of the problems and exercises in this book are intended to scaffold and maximize students' opportunities to internalize a variety of representations and to help them recognize the common structural elements of the mathematical ideas behind the symbols used.

Organization

The book is organized into six chapters that are aligned with the vision and spirit of the standards of the National Council of Teachers of Mathematics:

- Numbers and Operations
- Decimals and Percents
- Fractions, Percents and Ratios
- Algebra
- Geometry
- Probability and Statistics
- Appendix: Pictorial Templates

Each chapter contains a collection of tasks that focus on the development of conceptual understanding and the making of meaningful connections across a wide range of concepts between the chapters. In addition, each chapter includes master frames that allow the teacher to create their own tasks using the activities found within each section. Master frames have "(M)" in their title.

Chapters are organized according to conceptual difficulty. The latter part of each chapter contains the most difficult problems and exercises. The introductory exercises in each chapter prepare students to work with the pictures and the processes used in the more advanced problems.

Teacher notes are included at several strategic points throughout the book. These notes give teachers some guidance on the main points to keep in mind as they plan how to introduce a particularly important concept. Depending on the topic, teacher notes may also suggest the most effective manipulatives or step-by step examples on a complex problem or procedure.

The Appendix contains thirty-nine pictorial masters that can be used in a wide variety of ways to engage students in exploration, to practice additional conceptual development, or to extend the tasks from the previous chapters. At the end of this appendix you will find eighty different sample suggestions on how to use these graphic masters.

A key feature of this book is the purposeful and systematic treatment of transformations and translations of the various representations used. This book systematically engages students in translating and transforming the key mathematical concepts across and within a variety of representational systems.

Transformations and Translations

Researchers have identified five different representation systems used in the teaching and learning of mathematics:

1. Experience-based – or real world problems where their context facilitates the solution of a wide range of mathematical problems;

2. Manipulative or concrete models – like Base-10 blocks, counters, etc., where the built-in relationships within and between the models serve to represent mathematical ideas;

3. Pictures or diagrams – figures that may represent a mathematical concept or a specific manipulative model, such as the ones used throughout this book;

4. Spoken languages – i.e. the teacher saying the number one hundred thirty-two is quite different from the teacher writing the number 132 on the board for students to see;

5. Written symbols – these can include numbers, regular English sentences or more specialized languages, such as mathematical expressions, i.e. $x + 2$

Transformations and translations are processes that change how mathematical concepts are represented within and across representations.

Transformations change how a concept is represented within the same representation system. For example:

Changing ¼ to 0.25 is a transformation within the same symbolic representational system (numerical representation within written symbols).

On the other hand, translations change how a concept is represented by changing the representational system used. For example:

Asking students to identify the ratio of shaded to non-shaded rectangles among the pictures below require students to take a pictorial representation and create a matching symbolic one, namely 3:5 as shown in the image.

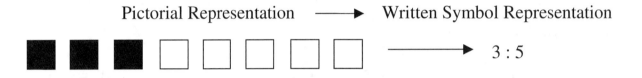

The book's pictorial representations use on a common set of structures across several concepts. For example, the structure of a subdivided rectangular figure is used throughout the book to conceptualize whole numbers, multiplication, fractions, decimals, percents, ratios, and expressions. By using the same structural pattern across various conceptual domains, students are able to build stronger and more efficient connections between the pictorial symbols used and the mathematical ideas embedded in the context of the problems.

Figure 1, in the next page, shows an example of the typical transformations done with the fraction ½. Whether we are talking about getting a 50% discount on an item at the store, getting ½ a cookie, coloring 12 out of 24 squares, or giving $0.50 in change, students should recognize "half" in its many different numeric forms. On the other hand, Figure 2 shows an example of the typical translations done in the context of ratios. Using the models and exercises in this book, students should be able to recognize the ratio of 3:2 in the standard numeric form, in pictures, in real life situations, in word problems, etc.

I invite the reader to take some time to consider the following question while analyzing figures 1 and 2: How many possible transformations are possible based on figure 1? How many possible translations are possible based on figure 2?

Figure 1
Transformations
To Build a Strong Conceptual Understanding
Build Bridges Between Multiple Representations With Transformations

Example 1

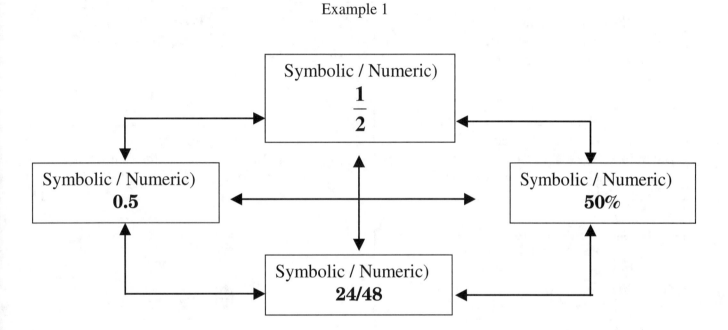

Transformations change the representation of a mathematical concept, but maintain the same representational system. In this example all four representations are symbolic/numeric.

Figure 2
Translations
To Build a Strong Conceptual Understanding
Build Bridges Between Multiple Representations With Translations

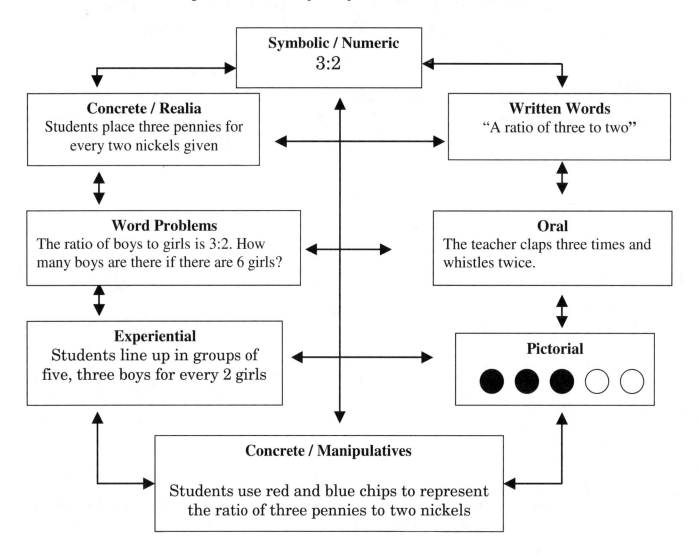

On fundamental, building-block concepts, such as proportional reasoning, expressions, and equations, teachers should spend extra time on how to translate and transform the key ideas among the eight different representational systems shown above. The extra time spent on these concepts, even at the expense of "covering" other material, will significantly improve students' conceptual understanding and overall achievement. (Figure 2 was sdapted from Lesh, 1979)

What is the main difference between translations and transformations and why are these processes so crucial in the teaching and learning of mathematics?

Both translations and transformations are processes that change how a particular mathematical idea is represented. The difference between them is similar to the differences between traveling by road, by air or by sea. Let's take someone who goes from point a to point b using a Volvo, then moves from point b to point c using a Honda. He has changed the way they traveled, from a Volvo to a Honda, but he maintained the same method of transportation: he traveled in both instances by car. Similarly, transformations change how a concept is represented, but they maintain the same representational system:

$$\frac{1}{4} \textit{ transformed to decimals is 0.25}$$

While the format of the number ¼ changed from fractions to decimals, both ¼ and 0.25 are numeric representations – just like both Hondas and Volvos are cars. Thus, transformations maintain the same representational system.

On the other hand, someone may go from Los Angeles to New York by air and come back from NY to LA by car. The actual method of transportation changed from a plane to a car. This is what happens with translations. Not only is the appearance of a mathematical idea or concept changed, but the actual representational system is also changed. Algebraic expressions are a good example of this. Here is one:

"Three more than some number"
translated to a mathematical expression is
"x + 3"

The original statement, "three more than some number", is given in standard English words. This statement was translated into a mathematical expression, or into a symbolic representation, x + 3.

Another example of translation is the use of base-10 blocks to represent numbers. A student can use the tangible manipulatives to construct the number 123. When a student is asked to construct the number 123 with the manipulatives, they are being asked to translate the standard numeric representation of the number 123 into a concrete[1] representation of the number 123 with the base-10 blocks.

Why should teachers make thoughtful and purposeful decisions about the types of representations, translations and transformations with which they engage students?

Let's stick with our transportation analogy: some means of transportation will get students from point a to point b faster, with less hassle. Others will provide students with more opportunities for sightseeing, for exploring, for generating lasting memories and for making more meaningful connections. Similarly, the way a teacher represents a given mathematical idea will have a great impact on the types of meanings, understandings, and connections students make. Again, a fraction-based example illustrates this idea well:

During the fourth and fifth grades, my teacher taught my classmates and me how to multiply two mixed fractions, such as 1 ½ times 4 ½ by following these steps:

- Change the 1 ½ to an improper fraction. To do so, multiply the whole number (1) by the denominator (2) and add it to its numerator (1). In our example, this gives us 1 x 2 + 1 = 3. Thus, (3) is the new numerator of your first fraction. Keep the same denominator (2). Thus, the new fraction is 3/2.

- Now change the 4 and ½ to an improper fraction. Again, the process is as follows: multiply the whole number (4) by the denominator (2) and add it to the numerator (1). This gives us 4 x 2 + 1 = 9. Thus, (3) is the new numerator of the second fraction. Keep the same denominator (2). Thus, the new fraction is 9/2.

- Multiply the numerators, then multiply the denominators. Your new fraction is 27/4.

- Is the numerator greater than the denominator? If yes, continue with the next step, otherwise, simply reduce the fraction. In our example, the numerator (27) is larger than the denominator (4), so we go to the next step.

---

[1] The base-10 blocks are considered here concrete in the sense that they are tangible objects that one can manipulate. However, base-10 blocks are symbolic in nature as they are representing the idea of number. Realia are concrete objects that are not representing anything other than what they are, for example, candies.

- Since the numerator is larger than the denominator, divide the numerator (27) by the denominator (4). That gives us a quotient of 6, remainder 3.

- Your final answer, a mixed number, will have the quotient (6) as the whole number, the remainder (3) as the numerator and the dividend (4) as the denominator. Thus, your final answer is 6 and ¾

This procedural, symbolic or abstract representation of the procedure to multiply mixed fractions made very little sense to me and to my classmates. It did not help us understand what multiplying two fractions meant. You either memorized the steps in the right order, or you did not get the right answer. But whether you got the right or the wrong answer, this way of representing the process of multiplying fractions did not promote conceptual understanding.

Compare the above process with a pictorial approach. A pictorial representation of 1 ½ x 4 ½ is based on the understanding of multiplication as "the repetition of something". That is, 1 ½ x 4 ½ is to be understood as 1 ½ groups of 4 ½ or repeating 4 ½ one and one half times. So, to do this multiplication pictorially, there are two steps:

- Draw $4\frac{1}{2}$ squares.

This represents 4 ½ repeated once

- To repeat 4 ½ one-half more time, draw exactly half of 4 ½ under the first set of squares.

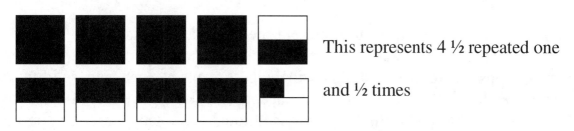

This represents 4 ½ repeated one and ½ times

Adding the pieces tells us that $1\frac{1}{2}$ times $4\frac{1}{2}$ is $6\frac{3}{4}$.

Most people who were not regularly taught with pictorial representations, at first, struggle with translating the symbolic representation of mixed fraction multiplication to a pictorial representation. But after a little practice, this way of representing fractions opens a new world of understanding previously unattainable to many students through the standard numeric representation.

In addition to helping students make more sense of the mixed-number multiplication, pictorial representations also opens a new window to the making of connections between multiplication and the concept of area.

Students who have gained a deep conceptual understanding of the multiplication of mixed numbers are able to translate the standard numeric representation of mixed fractions to a pictorial representation. They are also able to translate word problems to pictures, and word problems to symbols. In fact, teaching for conceptual understanding involves engaging students in at least twelve different translations and transformations.

Anyone teaching mathematics, by the very nature of mathematics, engages students with translations and transformations. Anytime we read aloud a number or a mathematical expression, we are translating a written symbol representation to an oral representation. However, what this book advocates for is to make translations and transformations a focal point of instructional planning and delivery; that is, to purposely plan activities that get students to translate and transform mathematical concepts within and across the five main representational systems; to monitor and become aware of the types of translations that we regularly use or tend not to use when we teach a given mathematical concept; and to view translations and transformations as pivotal instructional tools to help students develop strong conceptual understanding of mathematical concepts.

So what are the best ways to represent the key mathematical ideas we have to teach in order to facilitate students' conceptual understanding? What types of translations and transformations will help students deepen and solidify their understanding of key mathematical concepts and skills? The rest of this book is devoted to answering these questions through its many carefully designed models and exercises.

*"You don't understand anything
until you learn it more than one way."
- Marvin Minsky*

Methodology

There are two basic ways to use the materials found in this resource book. First, a teacher may choose to follow the sequence provided in the book. The book's sequence scaffolds instructional strategies to build a student's capacity to work independently through modeling, graphic organizers, working with partners, etc. A scaffolding technique used throughout the book is the inclusion of at least three equivalent worksheets for the key concepts: one for whole class work with the teacher modeling as needed, one for small group work, and one for individual practice. This built-in organizational scaffold helps the teacher to gradually release the responsibility for completing the tasks to the student.

The other way to use the book is to simply pick and choose the tasks that best fit the teacher's program. I recommend teachers continue to use the whole class, small group and individual sequence, even if they skip certain sections of the book as a whole. I also recommend, regardless of the order the materials, that students gradually move from the more concrete representations of a concept to the more abstract ones.

Tips On How To Use This Book Effectively

- ❑ **Teacher-practice**. Teachers should do the exercises before engaging students with them. Most of us educators were taught mathematics through symbolic manipulations, and learned to see mathematics as a set of rules to be memorized and applied. The problems in this book focus on developing meaning, understanding and connections. Doing the first couple of exercises on each page will provide teachers with the background needed to effectively support their students.

- **Concrete materials.** Teachers should use concrete materials in conjunction with the pictorial and symbolic representations used in this book. In particular, students will greatly benefit from the use of base-10 blocks and algebra tiles along with the pictorial tasks and activities found in this book. If you do not have these manipulatives, use the base-10 and algebra tile templates I have included in the appendix. In addition to using concrete materials, I highly recommend the classroom use of the virtual manipulatives available for free at Utah State University's website (**http://nlvm.usu.edu/en/nav/vlibrary.html**). This excellent site allows teachers and students to work with pictorial representations of most of the commercially available manipulatives. The teacher can use the site to model the use of the manipulatives as students learn to use the actual concrete materials. After working with the concrete materials, students can also work directly with the applets on the site. Students will gain a deeper conceptual understanding of key topics as they construct and move the various pictorial representations of the manipulatives within each applet.

- **Model how to explain one's reasoning.** Most students are not used to having to explain the process and the judgments they made to arrive at their solution. Teachers should model working a problem on the board, and then explain their reasoning, for example: "first I read the whole problem once, and then I re-read it, stopping after each sentence and wrote down the key information. I drew a picture of the problem to try to visualize it better. At first, I thought…"

- **Assessments.** Pre and post assessments should include the type of pictorial exercises presented in this book. Often, we tend to use the pictures and manipulatives during instruction, but somehow we end up assessing students only on the symbolic representations and procedures. Make sure your assessments align with the type of instruction and the type of problems students worked-out prior to the assessment.

- **Student presentations.** Teachers should ask students to share how they solved a task with the rest of the class. In order for students to feel comfortable sharing with the class, model making mistakes and show how you want students to react to them. Emphasize that mistakes are stepping-stones to the eventual correct solution. This helps to create an environment that rewards participation and supportive feedback.

- **Continuous connections between representations.** Students should continually be encouraged to make connections among a variety of representations. For example, whenever students are sent to the board to solve a problem, one student should do the work pictorially, another symbolically, another with words, and another with manipulatives on the overhead.

- **Use model-generative language.** The language we use while teaching mathematical concepts can serve to develop conceptual understanding or to retard it. For example, students should be taught to see ¼ as "one out of four equal parts", and 3 x 2 as either three groups of two or as two repeated three times. This type of language is modeled in the exercises in this book. Students who become proficient at reading the pictures while using such model-generative language will find it easier to develop a stronger conceptual understanding than those who simply read the pictures as "one -fourth" or "three times two."

- **Use an Interactive Learning Wall.** An Interactive Learning Wall is a dynamic bulletin board that calls for students to act or make a decision. For example, let's say a teacher is teaching equivalent fractions. The Interactive Learning Wall would have five or six different fractions, such as ⅓, ¼, etc. The teacher would prepare strips of paper containing equivalent fractions in different types of representations, for example, two sixths, a picture showing three squares, one of which is shaded, 0.25, and 25%. The teacher would place these strips in a paper bag. As students enter the room they each pick one strip from the bag. Students then go to the Interactive Learning Wall and tape their strip under the appropriate matching equivalent fraction. For more details on the learning wall see page 91.

- **Wait time.** When asking questions, teachers should commit to waiting longer whenever the task calls for students to perform either a translation or a transformation between representations. After at least 8 to 10 seconds of wait time, teachers should ask probing, open-ended questions such as "What can you tell me about the pictures shown in the example?"

- **Feedback on answers.** Teachers should not immediately correct a student who comes up with a wrong answer. Instead, they should ask the student to explain their reasoning, and involve the whole class in thinking about the solution offered. By using strategies such as asking the class to show thumbs up if they agree thumbs down if they disagree with the student's solution, the teacher can engage the whole class in thinking about a problem. Teachers should use similar strategies when the students give the right answers. By doing so, students will perceive these probing questions as invitations to explain, clarify and examine one's thinking, and not as clues to a wrong solution.

- **Master frames (M).** The book includes several master frames for key exercises. These pages have (M) in the title. They are meant for the teacher to fill in with their own questions using the previous exercises as their model.

- **Master graphic tools.** The last section includes master graphical resources that can be used for literally hundreds of tasks to help students develop conceptual understanding. I have included eighty sample tasks illustrate how one might use these graphical resources.

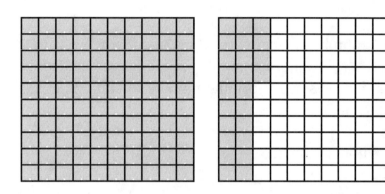

Number And Operations

Number and Operations

What Will Students Be Conceptualizing/Practicing?

- Place value: translating numbers across numeric, pictorial and written representations

- Adding whole numbers

- Constructing numbers: transforming numbers within pictorial and numeric representations

- Subtracting whole numbers

- Basic addition facts

- Adding integers

- Understanding multiplication

- Translating multiplications from pictorial to written, and numeric representations

- Transforming multiplications within written representations

- Translating multiplications from written representations to pictorial and numeric representations

- Transforming multiplications within numeric representations

- Translating multiplications across pictorial, numeric and written representations

- Recognizing multiplication as the repetition of something (a quantity)

- Creating the basis for multiplication word problems

- Multiplication models

- Recognizing the operation of multiplication is being used within word problem descriptions

- Translating multiplications from numeric to pictorial and from pictorial to numeric representations

- The box multiplication method: using the distributive property to simplify multiplications

- Translating pictorial representations of multiplication and division to written and numeric representations

- Prime factorization with factor trees and factor ladders

- Reading number lines

- Problem solving

- Generalizing

Pictorial Mathematics · Number and Operations

Teacher Notes
Place Value - Whole Numbers

Begin by pre-assessing what students know already about place value. Some sample questions may be:

- Write the number five hundred six and underline the number in the tens' place

- Write a number that has 4 in the tens' place, 8 in the hundreds' place, and 3 in the ones' place

- What is the difference between a digit and a number?

- What is the difference in the value of the digit 7 in the number 74 and in the number 57?

- You are cashing a $780 check and ask the bank teller to give you the entire amount using only ten-dollar bills. How can you use your knowledge of place value to know the number of ten-dollar bills that you should get?

- Introduce the Base-10 blocks to your students. Have them count from 1 to 10 using single units. How do then single units compare to one ten-unit bar?

- Have students construct the number twelve with single units, and then with one bar and two units. Which of these constructions matches the way we normally write the number twelve?

- Have students construct several two-digit numbers using the base-10 blocks. Make sure students are given the numbers in different ways: read the numbers to them, write the actual number in standard form (24), and/or write the number in words (sixty-seven).

- Construct different numbers with the overhead base 10- blocks and have the students read them, write them in standard form, and write them in words.

1 = □   10 = ▯   100 = ☐

Introduce it.

- After some discussion on the most effective ways to represent numbers pictorially, introduce the format presented above as the official class representation for the base-10 blocks.

- Lead students to work on translating the representation of various numbers, from symbolic to pictorial, from pictorial to symbolic, from words to pictures, etc., using the exercises on pages 41-43. Whenever possible, do not simply workout the example given on each worksheet. Rather, ask students to analyze the example and describe what they see to their partners, their group, or the entire class. Use page 44 to make up your own practice or homework worksheets.

Adding Whole Numbers. Introduce the idea of combining or adding numbers using base-10 blocks with this question. What is the most effective way to add 34 and 37 using the base-10 blocks? Have students go to the overhead and show their procedures for constructing the sum of 34 and 37 using the overhead blocks.

- Give students several pairs of numbers. Each pair should have one number written in standard form and one written pictorially. Ask students to add the numbers. For example:

    243 +   ▯▯ ☐ ☐ ☐

- Ask them to explain the easiest way to find their sum.

- Lead students through practice in adding numbers pictorially using the worksheets on pages 45-47. You will notice that there are several exercises in the form of addition that really call for subtraction. For example:

    _____ +   ▯ ☐ = 231

- Make sure that students have the actual manipulatives available for them to use throughout their learning experience. Mixing questions that require students to use manipulatives with questions that make them optional will make it easier for students to make their own choices as to whether to use that manipulatives in a given situation. Whenever possible, ask different students to show how they solve a problem with manipulatives, pictures, and/or standard symbols.

Subtraction of Whole Numbers. Whenever possible, give students time to learn how to subtract numbers with the actual base-10 blocks, before asking them to subtract their pictorial representations.

- Pair up students. Ask that one student work only with the manipulatives while the other works only with pencil and paper writing the numbers in standard form. After 5 or 6 problems, have them switch roles.

- (10 – 1) The student with manipulatives should get one bar of 10 units and then must find a way to take one unit way from this bar. To do this, they will have to "exchange" the bar of 10 units, for 10 single units. DO NOT use the term "borrow". One never borrows in subtraction, rather, one exchanges a unit of larger value for 10 units of the next lower value.

- (23 – 6) Have different students show their procedure for doing this subtraction with base-10 blocks. Stress the process of exchanging units. Be aware that many students have been taught to think and name the process of exchanging as "borrowing". Ask them to discuss in their groups the difference between borrowing and exchanging.

- Lead students to solve the subtraction problems in pages 50-52.

- If you do not have a class set of base-10 blocks, make copies of the base-10 templates in the appendix. If possible, make the copies onto card-stock, they last longer and are easier to handle. Have students cut out two sets of that page and put them in a plastic zip-lock bag.

Pictorial Place Value – Whole Numbers (1)

☐ = 100   ▯ = 10   ▫ = 1

Symbolic	Pictorial	Symbolic/Words
134	☐ ▯▯▯ ▫▫ ▫▫	One hundred thirty four
243		
	☐☐☐ ▯▯▯▯ ▫▫▫▫ ▫▫▫▫ (5 squares, 4 bars, 8 dots)	
		Three hundred six
530		
	▯▯▯▯ ▯▯▯▯ ▫▫▫	
345		
		Sixty-four

Pictorial Place Value – Whole Numbers (2)

☐ = 100 | = 10   ▫ = 1

Symbolic	Pictorial	Symbolic/Words
	☐ | | ▫▫▫▫ ▫▫▫▫	
234		
	☐☐ ☐☐☐ |||| ▫▫▫ ▫▫▫	
		Four hundred twenty
316		
	☐ ||||||	
200		
		One hundred twenty nine

Pictorial Place Value – Whole Numbers (3)

☐ = 100   ▯ = 10   ▫ = 1

Symbolic	Pictorial	Expanded form
126	☐ ▯▯ ▫▫▫ ▫▫▫	1(100) + 2(10) + 6 (1)
214		
	☐☐ ☐☐☐ ▯▯▯ ▫▫▫ ▫▫▫▫	
		3(100) + 4 (1)
230		
	☐ ▯▯▯▯▯▯ ▫▫▫	
		0(100) + 4(10) + 3 (1)
306		

Pictorial Place Value – Whole Numbers (M)

☐ = 100   ▯ = 10   ▫ = 1

Symbolic	Pictorial	Symbolic/Words

Adding Whole Numbers (1)

☐ = 100 | = 10   ▫ = 1

If column 1 + column 2 = column 3, draw the missing number, then write it.

Expression 1	+	Expression 2	=	Expression 3
☐ ||		☐ ||| ▫▫▫		☐☐☐☐ ||||||
		☐ ||||| ▫▫▫▫▫		☐☐☐☐ ||||||
☐ || ▫▫▫		☐ ||||| ▫▫▫▫▫		☐
☐				☐ ||| ▫▫
|| ▫▫▫▫▫				|| ☐ ▫▫

Adding Whole Numbers (2)

☐ = 100 | = 10   ▫ = 1

If column 1 + column 2 = column 3, draw/write the missing number.

Expression 1	+	Expression 2	=	Expression 3
240		☐ ||| ▫▫▫		
		☐ ||| ▫▫▫▫▫		321
		157		☐☐☐ || ▫
284		☐ |||| ▫▫▫▫▫▫		
|| ▫▫▫▫▫		203		

Adding Whole Numbers (3)

 = 100 = 10  = 1

If column 1 + column 2 = column 3, draw the missing number.

Expression 1 + **Expression 2** = **Expression 3**

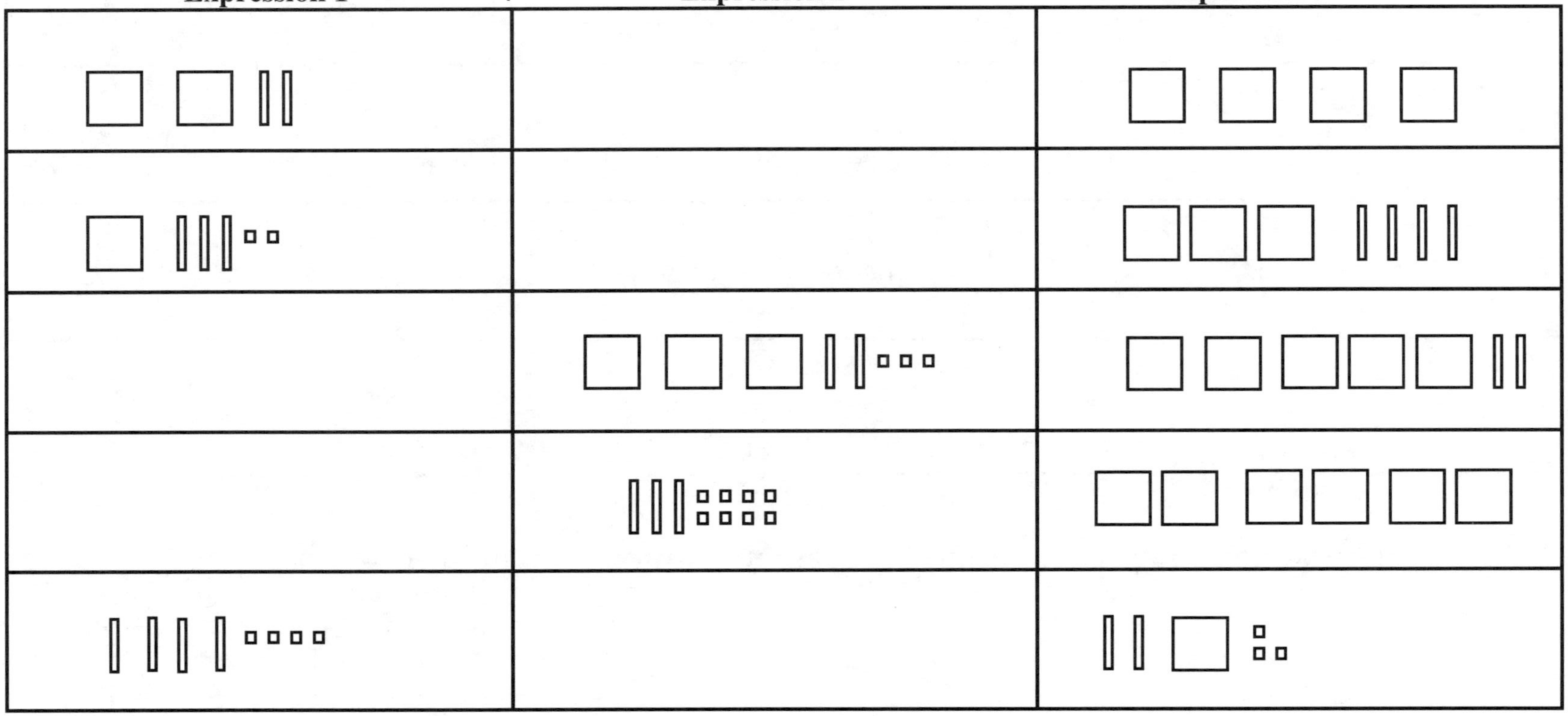

Pictorial Mathematics

Adding Whole Numbers (M)

☐ = 100   ∥ = 10   □ = 1

If column 1 + column 2 = column 3, draw the missing number, then write it.

Expression 1	+	Expression 2	=	Expression 3

Pictorial Mathematics   Number and Operations

Teacher Notes - Subtraction Examples

☐ = 100   ▭ = 10   ▫ = 1

The following are examples of pictorial subtraction without regrouping. Go over these examples with your students. If possible, have students construct them with base-10 blocks, then draw them, before doing the subtraction problems that follow.

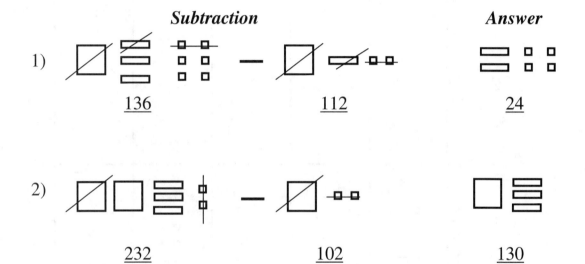

         Subtraction   *Answer*

1)   136   112   24

2)   232   102   130

The following is an example of pictorial subtraction with regrouping.

Subtraction

3)

Regrouping (not enough 10's)   *Answer*

      240   80   160

Subtracting Whole Numbers (1)

 = 100 = 10  = 1

If column 1 - column 2 = column 3, draw the missing number, then write it.

Expression 1	– Expression 2	= Expression 3 (show regrouping if needed)

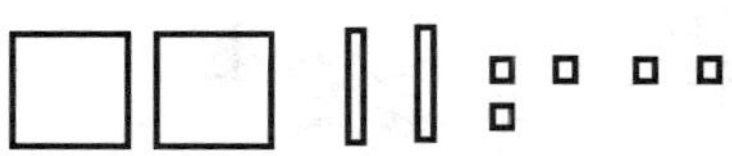		

Subtracting Whole Numbers (2)

☐ = 100   ▯ = 10   ▫ = 1

If column 1 − column 2 = column 3, draw the missing number, then write it.

Expression 1	Expression 2	Expression 3
☐ ▯	▯▯▯	
☐ ▯▯ ▫▫▫	☐ ▫▫ ▫▫	
☐ ▯▯ ▫	▫▫ ▫▫ ▫▫	
☐ ▯▯▯▯	▯▯ ▫▫ ▫▫	
☐ ▯▯▯ ▫▫	▯▯▯ ▫ ▫▫	

Subtracting Whole Numbers (3)

☐ = 100 | = 10   ▫ = 1

If column 1 − column 2 = column 3, draw the missing number, then write it.

Expression 1	−	Expression 2	=	Expression 3
☐☐☐ |	−	☐ ||| ▫▫▫	=	
☐☐ || ▫▫▫	−	☐ || ▫▫	=	
☐ |||| ▫▫▫	−	|||| ▫▫▫▫	=	
||||||	−	||| ▫▫▫ ▫▫▫	=	
☐	▫▫	−	||| ▫▫ ▫▫▫	=	

Thinking Through Basic Skills – Addition (1)

Using the digits 1 through 9 only once in each exercise, write one digit in each square so the operation makes sense.

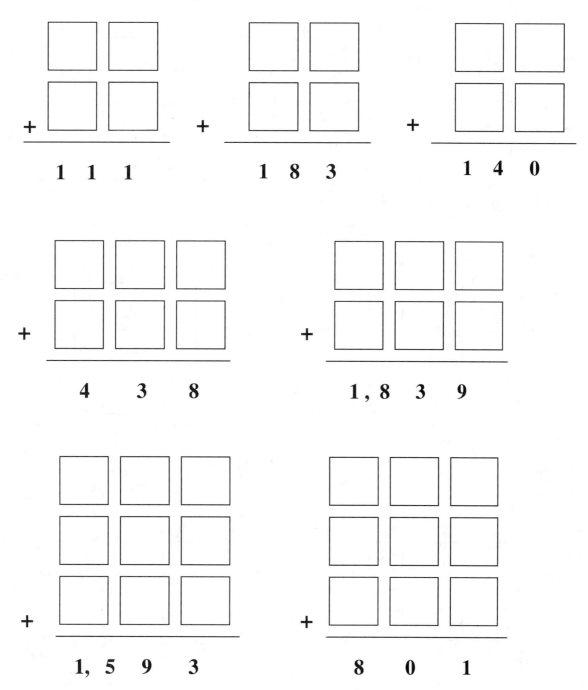

Pictorial Mathematics   Number and Operations

Thinking Through Basic Skills – Addition (2)

Using the digits 0 through 9 only once in each exercise, write one digit in each square so the operation makes sense.

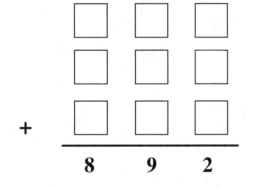

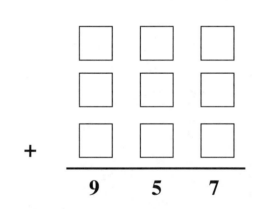

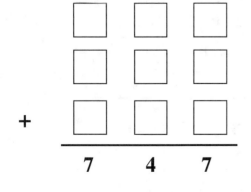

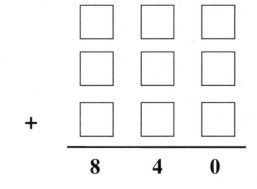

Thinking Through Basic Skills - Subtraction

Using the digits 1 through 9 only once in each exercise, write one digit in each empty square so the operation makes sense.

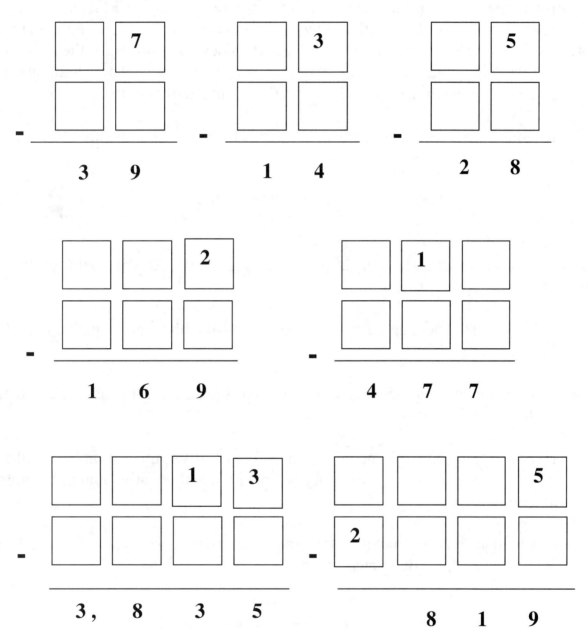

Teacher Notes
Integers

Base-10 blocks are a great resource to introduce the concepts of positive and negative numbers to your students. If you have sets of base 10 blocks, choose one color to represent the positive integers and another color to be the negative integers. If you do not have two-color base-10 blocks, simply copy the base-10 template in the appendix onto card stock and give two copies of this template to each student, one in white or a light color, and one in a darker color.

The first step is to introduce the value of the positive and negative base-10 blocks as follows:

1 = □   −1 = ■   10 = ▭   −10 = ▬   100 = ☐   −100 = ■

Once the values are established, the following exercises will help students understand the representation:

- Have students count aloud as they construct −10 with single negative units (-1, -2, -3……-10).

- Have students construct several number pairs with the same absolute value, such as −24, 24, -45, 45

- Have students add one negative unit to a positive unit. Emphasize that whenever one has the same number of positive and negative units, the units add up to zero.

- Ask students to construct the number 99 using only two blocks (positive 100 and negative 1), as illustrated here

- Ask students to construct several numbers using the least number of blocks. (i.e. 29, 48, 140, -39, - 338, etc.)

- Put students in groups of four. Assign each student a number from 1 to 4. Have all students numbered 1 construct a negative integer from −1 to −999 with the base-10 blocks or with pictures. Have all #2 students construct a positive integer from 1 to 999. Have all #3 students physically combine (add) these blocks. Have all #4 students represent these numbers and their sum in the standard form with pencil and paper. Students should rotate roles after three problems

- Lead students through the adding integer worksheets on pages 58-60.

- Play the integer card game. Put students in groups of four. Handout a pack of regular playing cards to each group. All the face cards are worth zero. All the black-number cards are positive integers worth the number on the card. All the red-number cards are negative integers worth the negative value of the number on the card. Each student gets five cards. Each player adds the numbers on his hand. Players can change any cards they wish once. The student with the highest total wins the hand. Each hand won gives the student one point. The first player to ten points wins the game.

- To practice multiplying integers you play a variation of the integer game above. Students get four cards instead of five. They choose which pair of cards produces the highest possible product. For example, a student with the cards - red 9, black 7, red 5, black 6 – should choose the red 9 and red 5 because when they are multiplied, they produce the highest possible product, namely, 45. The player with the highest product wins one point. Players must show their entire hand. If another player discovers someone did not choose the highest product in his or her hand wins an extra point.

Adding Positive and Negative Integers (1)

□ = 100 ■ = -100 □ = 10 ▬ = -10 ■ = 1 ■ = -1

If column 1 + column 2 = column 3, draw, then write the missing number.

# Adding Positive and Negative Integers (2)

□ = 100   ■ = -100   ■ = 10   □ = -10   ■ = 1   □ = -1

If column 1 + column 2 = column 3, draw the missing number.

Expression 1	+	Expression 2	=	Expression 3
-243		■ ■■ ■		121
		□ ■■■ □□		
		137		108
		■■■■ ■ □		
		301		

Adding Positive and Negative Integers (3)

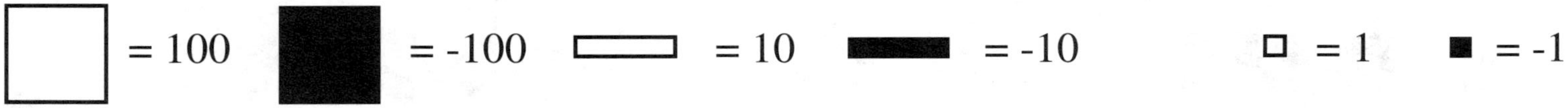

☐ = 100　■ = -100　☐ = 10　▬ = -10　☐ = 1　■ = -1

If column 1 + column 2 = column 3, draw/write the missing number.

Expression 1	+	Expression 2	=	Expression 3
		(■ plus 5 bars ▬ and 2 small ■)		328
-417		(2 ☐ squares and 1 bar ▬ and 4 small ☐)		
(4 ■ squares, 1 ☐ small bar, 2 small ■)		639		
		(2 ☐ squares, 4 bars ▬, 1 small ■)		236
		296		(1 ■ square, 2 bars ▬, 2 small ☐)

Multiplication

What Concepts and Skills About Multiplication Will Students Be Conceptualizing and Practicing?

- Develop conceptual understanding of the meaning of multiplication
- Recognize multiplicative structures
- Represent multiplication in a variety of ways
- Translate multiplication across several representations
- Identify the operation of multiplication within word problems
- Learn to multiply using several different procedures
- Learn the model-generative vocabulary that will facilitate connecting a situation with a learned multiplicative-model.
- Use the distributive property to simplify multiplications
- Factoring a trinomial to simplify multiplications
- Connect the pictorial model to words and symbols
- Connect the pictorial model to the standard multiplication procedure
- Summarize the different multiplication models
- Multiplying by $\frac{1}{10}, \frac{1}{100}$, 10, and 100
- Multiplying decimals
- Using factoring to multiply decimals
- Multiplying fractions
- Multiplying expressions

Teacher Notes – Multiplication

Few concepts have more to do with students experiencing success or failure in school mathematics than the concept of multiplication. A student who does not master the concept of multiplication by the third grade will fall behind very quickly. They will have problems learning division, fractions, percents, problem solving, area, expressions, equations, statistics, volume, and a long list of other key mathematical concepts and skills. Unfortunately, during elementary school, when students are beginning to be introduced to multiplication, this concept is often treated as a mere skill, as a procedure, as a set of facts to be memorized and recalled on demand.

In practical terms, multiplication is a concept to be understood as well as a skill to be mastered. Yes, students should end-up memorizing the basic multiplication facts, but this should be a by-product of a carefully crafted set of learning experiences that focus on the conceptual understanding of multiplication. There is no denying that a student gains a tremendous advantage when "they know their times tables." Teachers should require their students to memorize their times tables, but teachers should also use the strategies that have been proven most effective in helping students accomplish two key goals: 1) memorize the basic multiplication facts, and 2) gain a conceptual understanding deep enough to allow them to recognize when a situation calls for a multiplication, to properly apply multiplication in arithmetic and algebraic contexts, and to determine the reasonableness of a particular product.

This resource provides teachers with a conceptual approach to multiplication, along with a set of conceptual drills and exercises that will help students learn the meaning of multiplication, help them recognize when multiplication is embedded in a situation, and help them memorize the key facts that will allow them to have more time to focus on the problem solving process.

The Conceptual Language of Multiplication

The language we use to introduce and define multiplication has great implications for the type of conceptual understanding students develop about multiplication. Students should be introduced early to the idea of conceptualizing multiplication as *"the repetition of something"*:

- 2 repeated three times: 2 + 2 + 2, 3 groups of two or 3 x 2

- banana + banana + banana: 3 x banana

- half repeated three times: $\frac{1}{2} + \frac{1}{2} + \frac{1}{2} = 3 \times \frac{1}{2}$ or 3 halves $\left(\frac{3}{2}\right)$

The significance of thinking about multiplication as "the repetition of something" becomes more palpable when students are learning to multiply fractions:

$1\frac{1}{2} \times 1\frac{1}{2}$

Using the repetition idea, this multiplication can be read as:

$1\frac{1}{2}$ repeated once and then repeated half times

Using pictures to show $1\frac{1}{2}$ times $1\frac{1}{2}$ using this conceptualization becomes:

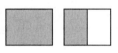   This shows $1\frac{1}{2}$ repeated once, that is $1\frac{1}{2} \times 1$

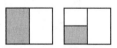   This shows $1\frac{1}{2}$ repeated half times ($1\frac{1}{2} \times \frac{1}{2}$)

So, using the idea of multiplication as the repetition of what is being multiplied, by adding the shaded pieces above we see that $1\frac{1}{2}$ times $1\frac{1}{2}$ is $2\frac{1}{4}$.

This resource provides teachers and students with a variety of flexible multiplication models that they can adapt to a wide variety of contexts. I strongly recommend that teachers spend as much time as possible engaging students with the multiplication conceptual models included in this section. It will pay off in a big way when they are learning about fractions and algebraic expressions.

Developing Multiplication Understanding (1)

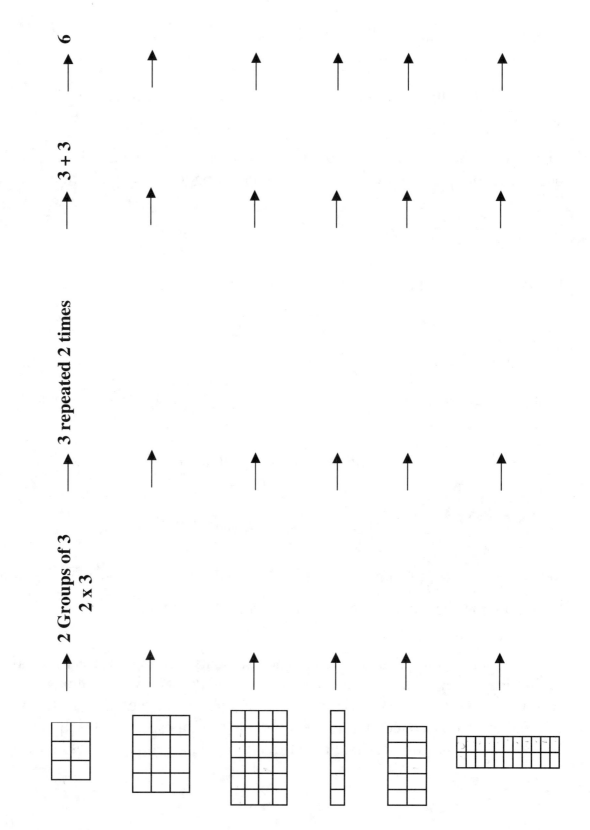

2 Groups of 3
2 x 3

3 repeated 2 times

3 + 3

6

Developing Multiplication Understanding (2)

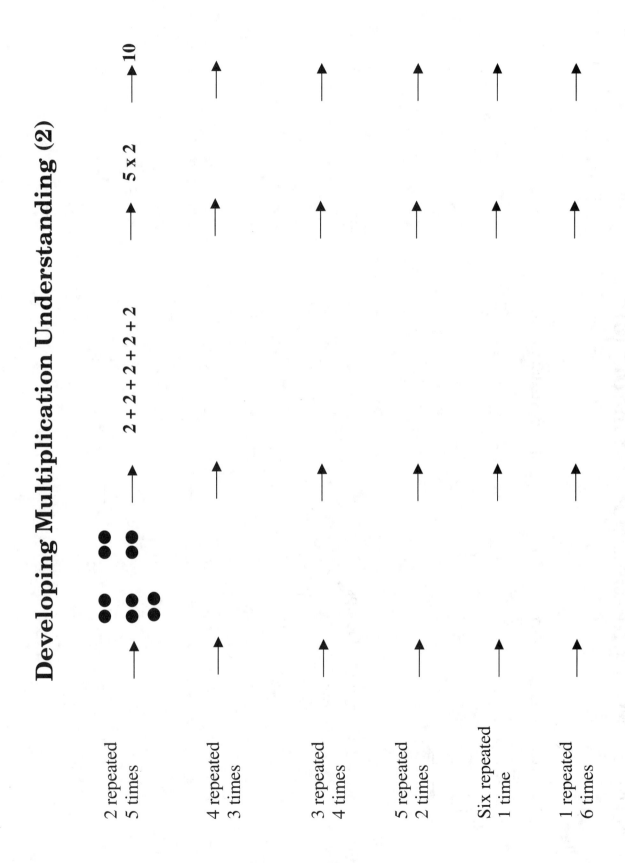

Developing Multiplication Understanding (3)

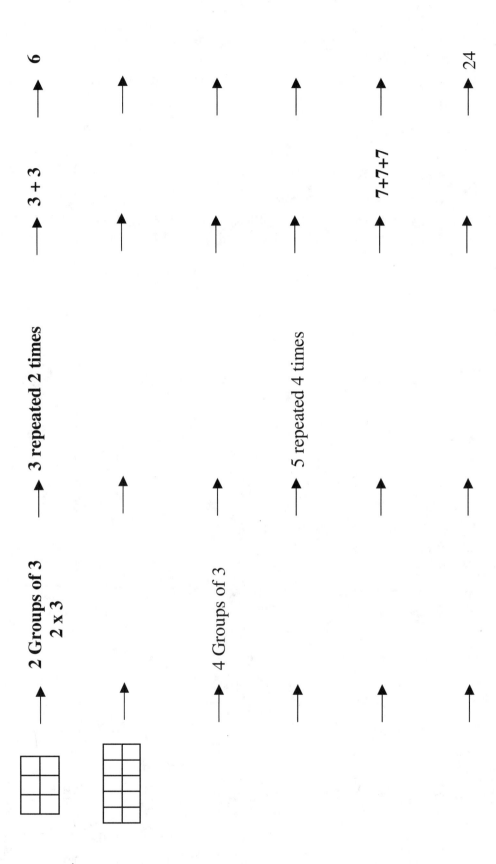

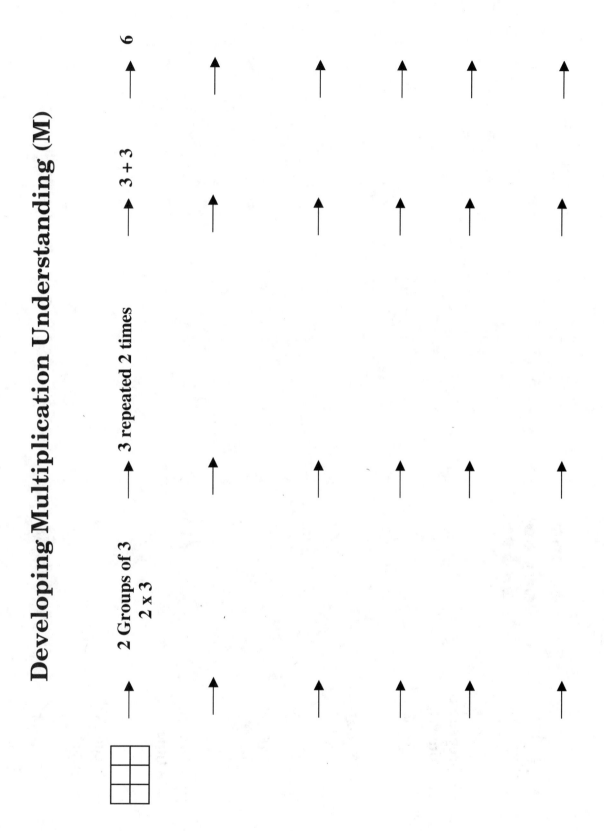

Developing Multiplication Understanding (2M)

5 repeated 3 times	→	●●●●● ●●●●● ●●●●●	→	5 + 5 + 5	→	3 x 5	→	15
___ repeated ___ times	→		→		→		→	
___ repeated ___ times	→		→		→		→	
___ repeated ___ times	→		→		→		→	
___ repeated ___ time	→		→		→		→	
___ repeated ___ times	→		→		→		→	

Multiplication Models

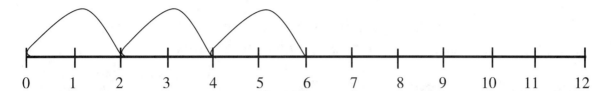

Three jumps of two spaces each represent 3 x 2. It can be read as three groups of two or as two repeated three times.

Five Groups of four objects in each group represent 5 x 4. It can also be read as four repeated five times.

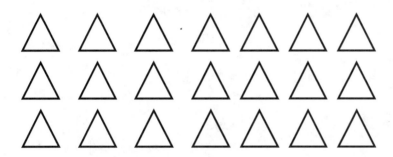
This three by seven array represents three rows of seven, 3 x 7. This array can also represent 7 x 3 (seven columns of three each)

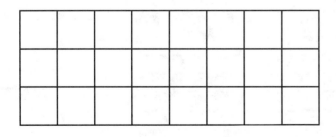
This is also an array, an eight by three array that represents 8 x 3 or 3 x 8.

Pictorial Mathematics   Number and Operations

Recognizing Multiplication (1)

Using the key word on the right, the picture, and the example, write the Multiplication being represented by the pictures.

Example:

Key Word

Legs

Three animals, four legs each : 3 x 4

Fingers

_____

_____

Wheels

_____

_____

Petals

_____

_____

Days

_____

_____

Pictorial Mathematics   Number and Operations

Recognizing Multiplication (2)

Using the key words on the right and the pictures below, write the multiplication being represented by the pictures.

Key Word

Eggs

Two Cartons with 12 eggs each

2 x 12 = 24 eggs

Cents

_____

_____

Cents

_____

_____

240 passengers   240 passengers   240 passengers

Passengers

_____

_____

71

Recognizing Multiplications 3

A game called Krugs is played with three groups of four. Students will not play the game, instead, they will read the rules of the game below and use the information to practice recognizing multiplications in a variety of forms.

At the start, every player takes turns throwing a pair of dice until someone gets six repeated four times. This can take a long time. A three by four rectangular grid is used to write down the scores each player gets. For example, if a player gets a two on one die and a four in the other, he gets two times four, or eight points. Players are in groups of four and the group's total is computed by adding the individual points from each other, for example, seven players with three points each gives the group 21 points. A player that has twice as many points as another at any time can challenge that player to a duo, and eliminate that player if he or she gets more points in four throws. Recording 6+6+6+6 means that a player threw four sixes in a row.

What terms are used to signify multiplication?	What multiplication?	Draw it as a grid

Linking the Standard Multiplication to a Pictorial Representation (1)

Example

Place value multiplication

```
   11
 x 12
 ----
    2
   20
   10
  100
 ----
  132
```

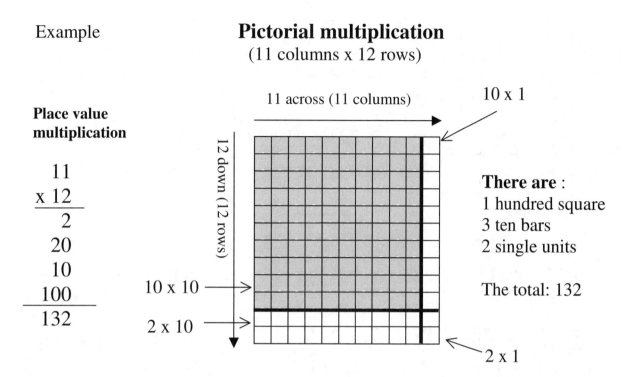

Pictorial multiplication
(11 columns x 12 rows)

There are:
1 hundred square
3 ten bars
2 single units

The total: 132

1) Explain how both the pictorial multiplication and the place value multiplication on the left of the picture, are similar or different than the standard way of multiplying 12 x 11.

2) Explain how the total 132 relates to the picture.

Introducing 2-Digit Pictorial Multiplications

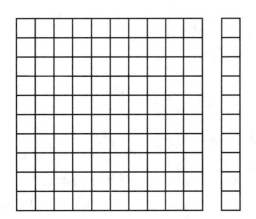

The grids on the left show a total of 11 columns (across) and 10 rows (down). This picture represents the multiplication 11 x 10 (11 columns times 10 rows)

The picture of 11 x 10 can be represented as:

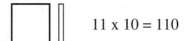 11 x 10 = 110

Use the example given above to fill-in the blanks

a)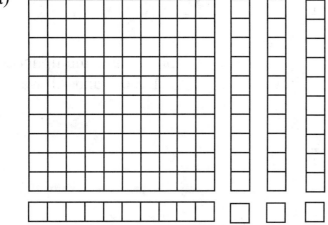

The grids on the left show a total of ____ columns across and ____ rows going down. This picture represents the multiplication

____ x ____

(____ columns times ____ rows)

The picture of ____ x ____ can be represented as:

b) = 12 x ____ = 132   c) ⬜||||| = ____ x ____ = ____

74

Pictorial Mathematics — Number and Operations

Base 10 Double Digit Multiplication (1)

☐ = 100 | = 10   ▫ = 1

Symbolic	Pictorial	Expanded Notation	Product
12 x 13		1 (100) + 5 (10) + 6(1)	156
12 x 14			
13 x 14		1 (100) + 8 (10) + 2 (1)	
13 x 15			
20 x 11			
21 x 11			
22 x 12			

75

Base 10 Double Digit Multiplication (2)

 = 100  = 10 ▫ = 1

Symbolic	Pictorial	Expanded Notation	Product
22 x 13	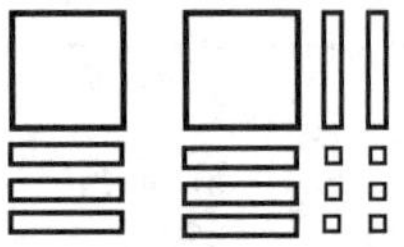	2 (100) + 8 (10) + 6(1)	286
12 x 15			
16 x 11		1 (100) + 7 (10) + 6 (1)	
13 x 14			
	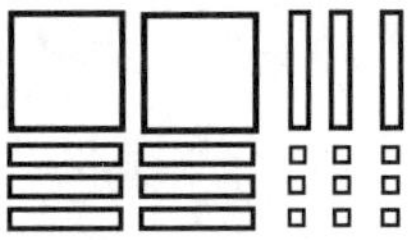		
	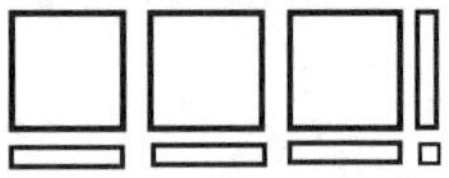		
31 x 12			
42 x 11			

Pictorial Mathematics   Number and Operations

Base 10 Double Digit Multiplication (M)

☐ = 100 | = 10   ▫ = 1

Symbolic	Pictorial	Expanded Notation	Product
12 x 13		1 (100) + 5 (10) + 6(1)	156

77

Pictorial Mathematics — Number and Operations

Box Multiplication (1)

Use example 1 to complete the box multiplications

1) 23 x 12

	20	3
10	200	30
2	40	6

```
  200
   30
   40
 +  6
 ----
  276
```

2) 32 x 13

3) 24 x 15

4) 62 x 21

Box Multiplication (2)

Complete the box multiplications and compute the total.

1) 14 x 22

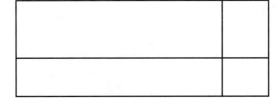

2) 34 x 11

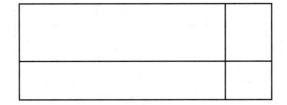

3) 54 x 25

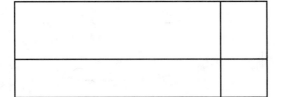

4) 42 x 21

Box Multiplication (3)

Complete the box multiplications, and compute the total.

1) 19 x 21

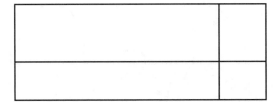

2) 33 x 11

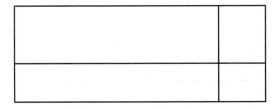

3) 52 x 23

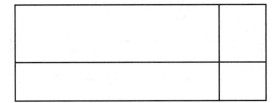

4) 43 x 31

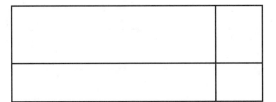

Pictorial Mathematics   Number and Operations

Understanding The Process
Of Multiplying And Dividing Operators (1)

Example:

| An arrangement of four squares. | Was multiplied 3 times. That is, was repeated 3 times. | The resulting arrangement was divided into 6 equal groups. The result shows the number of squares in one of these 6 equal groups. |

| 4 multiplied by 3 and divided by 6 is 2; in other words, (3 x 4) ÷ 6 = 2 |

Use the example above to complete the missing parts below

| An arrangement of _____ squares. | Was multiplied ____ times. That is, was repeated _____ times. | The resulting arrangement was divided into ____ equal groups. The result shows the number of squares in one of these ___ equal groups. |

| ___ multiplied by ___ and divided by ___ is ____; in other words, (___ x ___) ÷ ____ = |

Understanding The Process
Of Multiplying And Dividing Operators (2)

| An arrangement of _____ squares. | Was multiplied ____ times. That is, was repeated _____ times. | The resulting arrangement was divided into ____ equal groups. The result shows the number of squares in one of these ___ equal groups. |

___ multiplied by ___ and divided by ___ is ____; in other words, (___ x ___) ÷ ____ = ____

2)

| An arrangement of _____ squares. | Was multiplied ____ times. That is, was repeated _____ times. | The resulting arrangement was divided into ____ equal groups. The result shows the number of squares in one of these ___ equal groups. |

___ multiplied by ___ and divided by ___ is ____; in other words, (___ x ___) ÷ ____ = ____

3)

| An arrangement of _____ squares. | Was multiplied ____ times. That is, was repeated _____ times. | The resulting arrangement was divided into ____ equal groups. The result shows the number of squares in one of these ___ equal groups. |

___ multiplied by ___ and divided by ___ is ____; in other words, (___ x ___) ÷ ____ = ____

Understanding The Process
Of Multiplying And Dividing Operators (3)

1)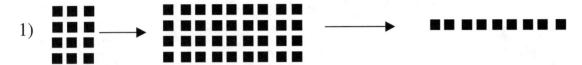

___ multiplied by ___ and divided by ___ is ___; in other words, (___ x ___) ÷ ___ = ___

2)

___ divided by ___ and multiplied by ___ is ___; in other words, (___ ÷ ___) x (___) =

3)

___ divided by ___ and multiplied by ___ is ___; in other words, (___ ÷ ___) x (___) =

4)

___ multiplied by ___ and divided by ___ is ___; in other words, (___ x ___) ÷ ___ = ___

Factor Trees and Factor Ladders (1)

Factor trees and factor ladders are two strategies to find the prime factors of a number. Use the example given to find the prime factorization of each number.

Factor Trees

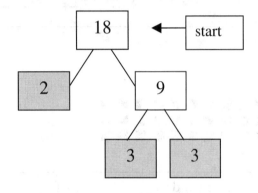

Prime Factorization: 2×3^2

Factor Ladders

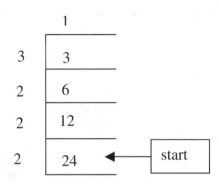

Prime Factorization: $2^3 \times 3$

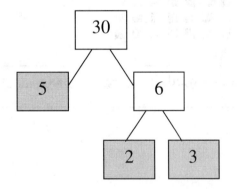

Prime Factorization:

Prime Factorization:

Factor Trees and Factor Ladders (2)

Factor trees and factor ladders are two strategies to find the prime factors of a number. Use the example given to find the prime factorization of each number.

Factor Trees

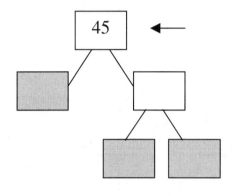

Prime Factorization:

Factor Ladders

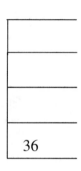

Prime Factorization:

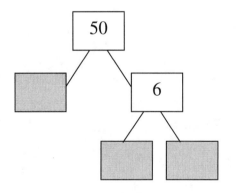

Prime Factorization:

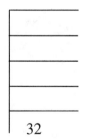

Prime Factorization:

Factor Trees and Factor Ladders (3)

Factor trees and factor ladders are two strategies to find the prime factors of a number. Use the example given to find the prime factorization of each number.

Factor Trees

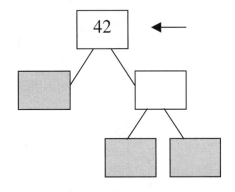

Prime Factorization:

Factor Ladders

63

Prime Factorization:

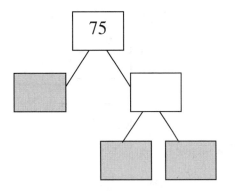

Prime Factorization:

100

Prime Factorization:

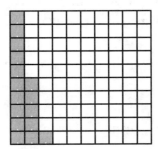

Decimals

Decimals

What Will Students Be Conceptualizing/Practicing?

- Translating decimals across numeric, pictorial, and written representations

- Translating positive and negative decimals from numeric to pictorial and from pictorial to numeric representations

- Transforming pictorial representations of decimals within numeric and pictorial representations

- Comparing decimal and fractions

- Estimating decimals

- Adding decimals

- Subtracting decimals

- Multiplying decimals

- Dividing decimals

- Connecting decimals, fractions and percent

- Translating positive and negative decimals from pictorial to numeric and written representations

- Adding positive and negative decimals

- Multiplying positive and negative decimals by positive and negative powers of 10.

- Factoring decimals using addition

- Factoring decimals using subtraction
- Using different representations of zero
- Transforming decimals to percents and fractions
- Proportional reasoning
- Estimating decimals
- Estimating percents
- Abstracting
- Generalizing

Teacher Notes – Decimals

The conceptual exercises in this section focus on four key processes:

1) Representing decimals with base-10 blocks

2) Representing decimals with pictures

3) Translating standard numeric representation of decimals to and from pictures and words

4) Transforming decimals to fractions, percents, and expanded form

I recommend that teachers begin by giving students the standard numeric representation (0.1, 0.04, 1.27, etc.) and the word representation (one tenth, four hundredths, one and twenty-seven hundredths, etc.) of simple decimals. Teachers then should model the construction of these decimals with the transparency base-10 block set and ask students to construct them with their own base-10 blocks.

Students should see all the different ways of translating and transforming decimals across a variety of representations. The following translation exercises should be used:

Given a decimal written in standard numeric form:

- Create a picture
- Construct the decimals with base-10 blocks
- Write the decimals in words
- Write the decimal in expanded form

These types of translation exercises should be repeated across each of the five basic representations:

Given these types of representation	Translate them to these types of representations
Numeric	Pictorial, words, concrete, oral
Pictorial	Numeric, words, concrete, oral
Words	Numeric, pictorial, concrete, oral
Oral	Pictorial, concrete, words, numeric
Concrete	Oral, pictorial, numeric, words

Working with Negative Decimals

I recommend that student' work with base-10 blocks of two different colors: one color to represent the positive decimals and another to represent the negative ones. If two-color base-10 blocks are not available, copy the base-10 block template in the appendix section onto two different colors of paper. Give students plenty of practice constructing the number zero by combining the same number of positive and negative units, tenths, and hundredths.

The Interactive Learning Wall

The interactive learning wall is a dynamic instructional strategy that can be used with any concept. The Wall helps to actively engage students and gives teachers a great way to quickly assess students' understanding of the material.

The Interactive Learning Wall can be a section of your bulletin board, blackboard or whiteboard. The teacher will use the board to involve the class in translating concepts and facts across a variety of representations. Lets say you have been working with your students on transforming decimals to fractions and percents, and on translating decimals across a variety of representations. The teacher would place the following types of number representations as headings on the wall:

a) ½   b) two thirds   c) 75%   d) ▢▯▯ ▫▫▫   e) other

The teacher writes or draws a variety of other representations for these headings on separate small strips of paper, folds them, and places them in a grab-bag. Examples of the type of representations for ½ that a teacher may write on these strips include:

1) Half

2) ▤

3) 50%

4) $\frac{1}{4} + \frac{1}{4}$

5) Six divided by twelve

Students reach into the bag and grab one strip. They must then go to the Interactive Learning Wall and tape their strip under the heading that matches the numeric representation of their strip. After several students have posted their strips, the teacher engages the class in assessing if all strips have been correctly matched.

Pictorial Decimal Representations (1)

☐ = 1   ▭ = 0.1   ▫ = 0.01

Symbolic Representation	Pictorial Representation
1.23	☐ ▭ ▫▫▫
2.06	
_____	☐☐ ▭ ▫▫▫▫
3.4	
_____	☐ ▭▭▭▭▭ ▫▫▫▫▫▫▫▫
4.21	
_____	☐ ▫
3.52	

Pictorial Decimal Representations (2)

☐ = 1   ▭ = 0.1   ▫ = 0.01

Symbolic Representation	Pictorial Representation
0.34	
3.20	
_____	☐ ☐ ▭▭▭
2.04	
_____	☐☐ ▭▭ ▫▫▫▫▫▫▫ ☐☐
0.21	
_____	☐ ☐ ▭▭▭ ▫
Twenty-two hundredths	

Pictorial Decimal Representations (3)

Symbolic Representation	Pictorial Representation
$1 - 0.1 - 0.1 + 0.01 = 0.81$	
$-2 + 0.01 = -1.99$	

$-2 + 1 + 0.1 + 0.01 = 1.11$	
	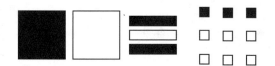
$1 + 1 - 0.1 - 0.01 = 2.09$	

Pictorial Decimal Representations (M)

☐ = 1   ▭ = 0.1   ▫ = 0.01

■ = -1   ▬ = -0.1   ▪ = -0.01

Symbolic Representation	Pictorial Representation

Pictorial Place Value – Decimals (1)

☐ = 1 | = 0.1 = $\frac{1}{10}$ ▫ = 0.01 = $\frac{1}{100}$

Fraction	Pictorial	Decimal	Words	Expanded
$1\frac{7}{10}$	☐ ||||||			
	☐☐ |||| ▫			
		0.36		
			Six and three hundredths	
				$4(1) + 3\left(\frac{1}{10}\right)$

Pictorial Place Value – Decimals (2)

□ = 1   ▯ = 0.1 = $\frac{1}{10}$ ▫ = 0.01 = $\frac{1}{100}$

Fraction	Pictorial	Decimal	Words	Expanded
$2\frac{1}{10}$				
		1.06		
				$1(1) + 8\left(\frac{1}{100}\right)$
		0.40		
			Seventeen hundredths	

Pictorial Place Value – Decimals (3)

□ = 1   ▭ = 0.1 = $\frac{1}{10}$ ▫ = 0.01 = $\frac{1}{100}$

Fraction	Pictorial	Decimal	Words	Expanded
$\frac{7}{10}$				
	(1 large square, 5 bars, 4 small squares)			
		0.3		
			Fifty-seven hundredths	
				$6\left(\frac{1}{10}\right)+5\left(\frac{1}{100}\right)$

Pictorial Place Value – Decimals (M)

☐ = 1   [] = 0.1 = 1/10   ▫ = 0.01 = 1/100

Fraction	Pictorial	Decimal	Words	Expanded

Pictorial Decimal/Fractional Representation 1

□ = 1   ▭ = 0.1 = $\frac{1}{10}$ ▫ = 0.01 = $\frac{1}{100}$

Pictorial Representation	Decimal Representation	Fraction	Word Representation
3 squares, 2 bars, 4 small squares			
2 squares, 5 bars, 4 small squares			
1 bar, 6 small squares			
4 squares, 3 bars, 6 small squares			
1 small square, 8 small squares			
2 squares, 5 bars			

Pictorial Decimal/Fractional Representation 2

☐ = 1   ▭ = 0.1 = $\frac{1}{10}$ ◻ = 0.01 = $\frac{1}{100}$

Pictorial Representation	Decimal Representation	Fraction	Word Representation
	1.34		
▭ ◻◻			
		$2\frac{2}{10}$	
	3.4		Three and sixty-two hundredths
			Twenty-four hundredths

Pictorial Mathematics — Decimals — 101

Pictorial Decimal/Fractional Representation 3

☐ = 1   ■ = -1   [] = 0.1 |] = -0.1   ▫ = 0.01   ▪ = -0.01

Pictorial Representation	Decimal Representation	Fraction	Word Representation
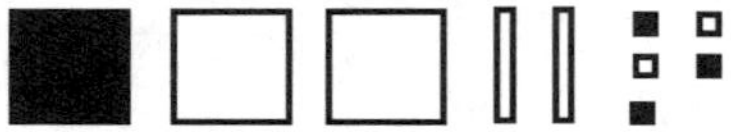 (large black square, 2 white squares, 3 horizontal bars, 3 small squares with 1 filled)			
 (large black square, 4 horizontal bars with 1 filled, 3 small squares with 2 filled)			
 (2 horizontal bars with 1 filled, 6 small squares)			
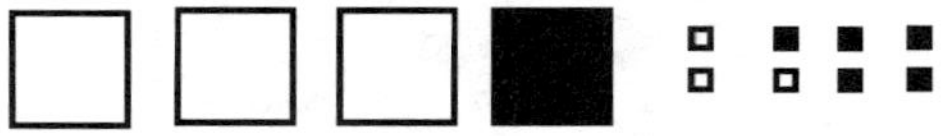 (3 white squares, large black square, 6 small squares with 3 filled)			
 (7 small squares with 3 filled)			
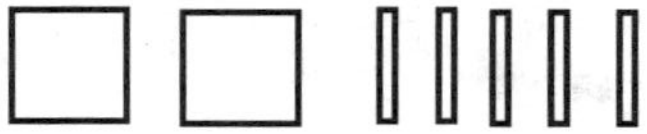 (2 white squares, 4 horizontal bars)			

Decimal Shadings (1)

Shade each figure to represent the decimal number given.

0.23 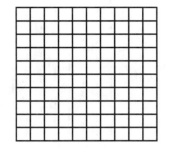   0.4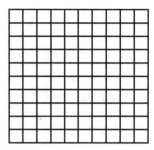

0.61 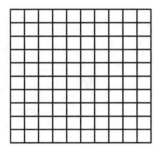   0.7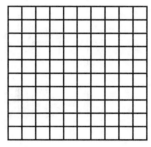

0.42 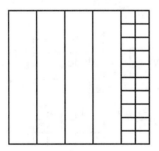   0.28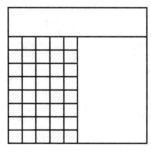

0.32 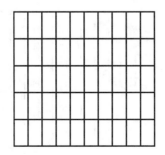   0.47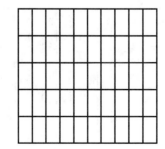

Decimal Shadings (2)

Shade each figure to represent the decimal number given.

0.30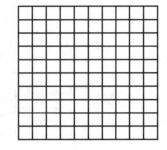

0.52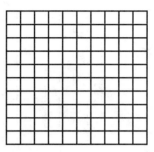

0.19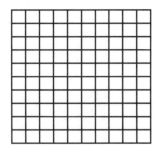

0.9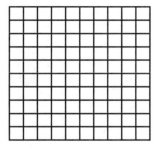

0.34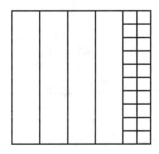

0.43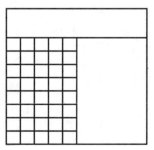

0.14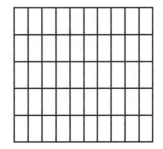

0.35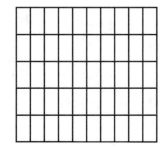

Percent Shadings (1)

Shade each figure to represent the percent shown.

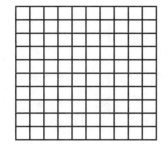 36%

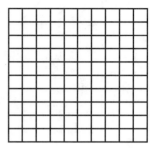 60%

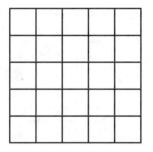 24%

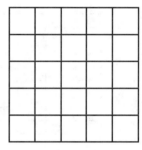 72%

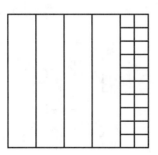 86%

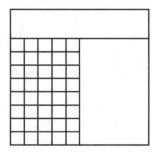 56%

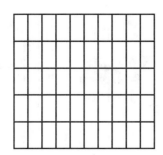 11%

0.43

Percent Shadings (2)

Shade each figure to represent the percent given:

40% 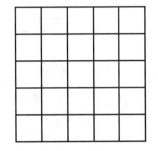   14%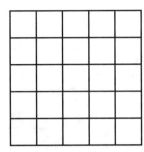

22% 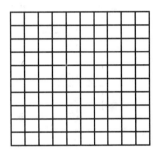   30%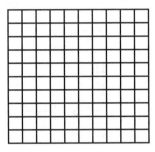

42% 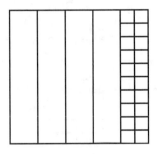   71%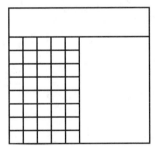

31% 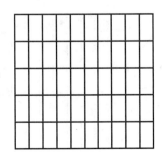   3%

Percent Shadings (3)

Shade each figure to represent the percent given:

30% 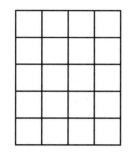   45%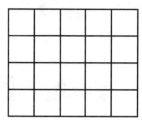

34% 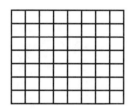   48%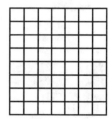

21% 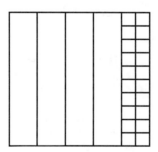   30%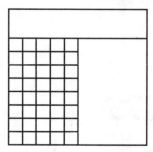

27%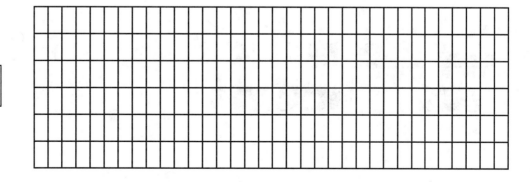

Pictorial Mathematics   Decimals

Estimating Decimals (1)

Underline the number that best estimates the number of shaded figures.

Example    3.1   2.1   <u>2.5</u>   3.0

1.    3.9   3.7   3.1   3.3

2.    4.2   3.5   3.7   3.9

3.    4.1   2.8   3.4   3.2

4.    2.4   3.2   4.1   3.7

5.    3.7   4.0   4.3   3.5

6.    5.0   4.7   5.5   6.0

Estimating Decimals (2)

Underline the number that best estimates the number of shaded figures.

Example   3.1   2.1   <u>2.5</u>   3.0

1.   3.8   2.9   3.1   3.5

2.   4.7   6.5   5.7   4.9

3.   3.1   2.8   2.4   3.75

4.   1.9   3.2   2.1   2.4

5.   3.4   2.9   3.8   2.5

6.   5.0   4.4   4.2   4.0

Estimating Percents (1)

Underline the number that best estimates the percent of the figures that is shaded.

Example  50% 75% 60% 25%

1. 5% 25% 30% 16%

2. 34% 65% 49% 40%

3. 27% 32% 35% 44%

4. 60% 74% 86% 67%

5. 27% 34% 42% 51%

6. 61% 51% 71% 56%

Estimating Percents (2)

Underline the number that best estimates the percent of the figures that is shaded.

Example 50% <u>75%</u> 60% 25%

1. 52% 28% 31% 37%

2. 24% 55% 36% 44%

3. 57% 71% 64% 82%

4. 80% 75% 66% 69%

5. 47% 36% 42% 31%

6. 31% 51% 44% 56%

Estimating Percents (3)

Underline the number that best estimates the percent of the figures that is shaded.

Example  50% <u>75%</u> 60% 25%

1. 49% 34% 61% 37%

2. 24% 55% 56% 41%

3. 47% 41% 58% 82%

4. 89% 75% 66% 81%

5. 49% 56% 62% 52%

6. 41% 61% 52% 54%

Pictorial Mathematics   Decimals

Comparing Decimals and Fractions (1)

Use fraction notation to list the following from least to greatest:

1. a) Four tenths   b) [pictorial: 3 tenths, 2 hundredths]   c) $\dfrac{38}{100}$ d) 0.308

2. a) One and 3 tenths   b) [pictorial: 1 whole, 4 hundredths]   c) $\dfrac{128}{100}$ d) 1.299

3. a) Fifty six hundredths   b) [pictorial: 5 tenths]   c) $\dfrac{29}{50}$ d) 0.59

4. a) Two and six tenths   b) [pictorial: 2 wholes, 8 hundredths]   c) $\dfrac{27}{10}$ d) 2.09

5. a) Eight hundredths   b) [pictorial: 3 tenths]   c) $\dfrac{3}{20}$ d) 0.2

6. a) Sixty four hundredths   b) [pictorial: 3 tenths, 2 hundredths]   c) $\dfrac{1}{4}$ d) 0.8

Comparing Decimals and Fractions (2)

Use fraction notation to list the following from least to greatest:

1. a) Six hundredths b) ▭▭▭ c) $\dfrac{8}{100}$ d) 0.07

2. a) One and 5 tenths b) 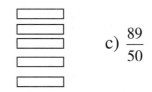 c) $\dfrac{148}{100}$ d) 1.09

3. a) Ninety two hundredths b) ▭▭▭▭ c) $\dfrac{89}{50}$ d) 0.69

4. a) Two and eight tenths b)   c) $\dfrac{270}{10}$ d) 2.09

5. a) Five hundredths b) ▭▭▭ c) $\dfrac{2}{10}$ d) 0.09

6. a) Fifty two hundredths b) ▭▭▭ ▫▫ c) $\dfrac{1}{4}$ d) 0.7

Comparing Decimals and Fractions (3)

Use fraction notation to list the following from least to greatest:

1. a) Two hundredths   b) [pictorial: 2 bars + 1 small square]   c) $\frac{3}{100}$ d) 0.208

2. a) One and five tenths   b) [pictorial: 1 large square + 5 small squares]   c) $\frac{120}{100}$ d) 0.130

3. a) Fifty six tenths   b) [pictorial: 5 bars]   c) $\frac{57}{100}$ d) 5.70

4. a) One and two tenths   b) [pictorial: 2 large squares + 1 small square]   c) $\frac{107}{100}$ d) 1.05

5. a) Sixty-one hundredths   b) [pictorial: 4 bars]   c) $\frac{3}{25}$ d) 0.13

6. a) Four hundredths   b) [pictorial: 2 bars + 1 small square]   c) $\frac{1}{2}$ d) 0.399

10 by 10 Shading (1)

If a 10 by 10 square is 1 unit, shade in the specified amount

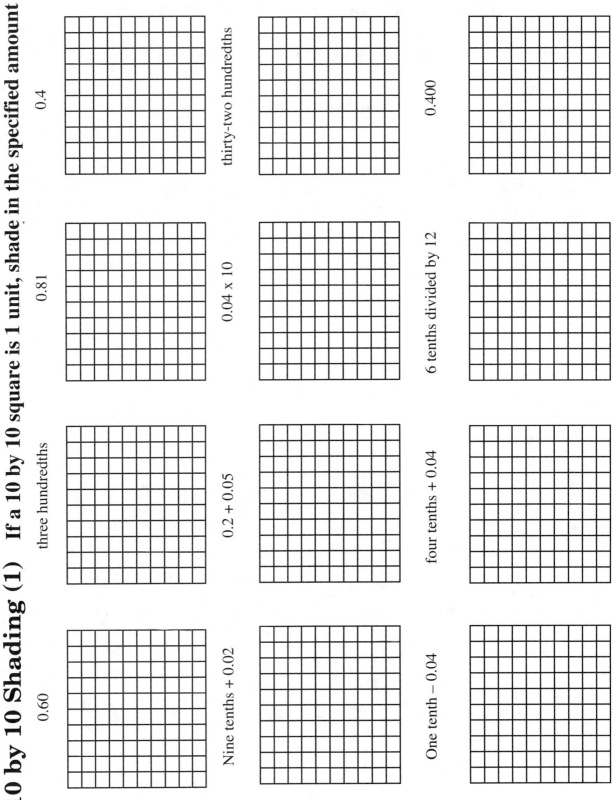

Pictorial Mathematics   Decimals

10 by 10 Shading (2)   If a 10 by 10 square is 1 unit, shade in the specified amount

$\frac{7}{10}$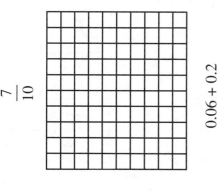

$0.06 + \frac{1}{20}$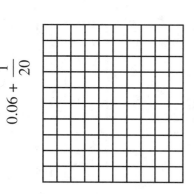

$\frac{4}{50}$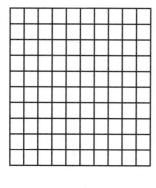

$\frac{18}{25}$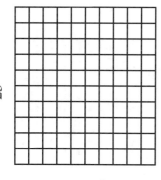

$0.06 + 0.2$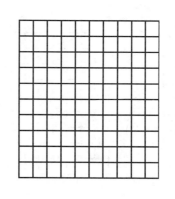

0.7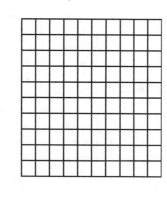

24%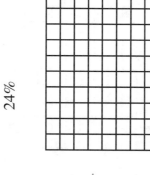

forty-two hundredths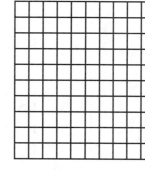

$0.08 \div 0.1$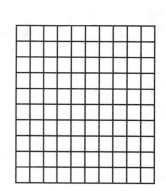

six tenths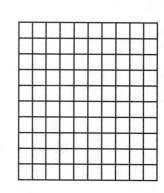

70%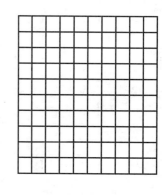

$9 \div 20$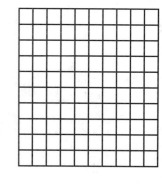

117

10 by 10 Shading (3)

If a 10 by 10 square is 1 unit, shade in the specified amount

0.18 divided by 0.4

fifty-two hundredths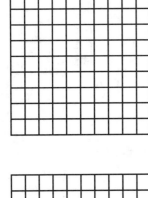

60%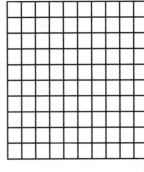

$17 \div 20$

0.09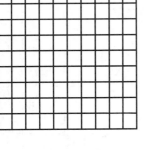

$9 \div 50$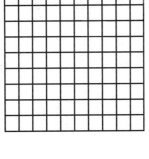

31%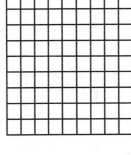

$0.04 \div 0.02$

$\dfrac{3}{10}$

four tenths + $\dfrac{1}{100}$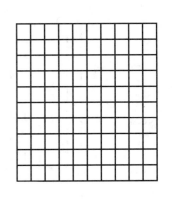

$0.7 + \dfrac{2}{50}$

$\dfrac{11}{20} + \dfrac{1}{5}$

Pictorial/Symbolic Decimal Addition 1

The number in the pictorial column plus the number in the decimal column should equal the number in the word column.

☐ = 1   ■ = -1   ▭ = 0.1   ▬ = -0.1   □ = 0.01   ■ = -0.01

Pictorial Representation +	Decimal Representation	=	Word Representation
	2.5		
	-0.84		
	-2.76		
	5.89		
	0.38		

Pictorial/Symbolic Decimal Addition 2

☐ = 1 ■ = -1 ▭ = 0.1 ▬ = -0.1 □ = 0.01 ■ = -0.01

The number in the pictorial column plus the number in the decimal column should equal the number in the word column.

Pictorial Representation +	Decimal Representation	=	Word Representation
	-1.4		
	-0.26		
	3.53		
	2.99		
	1.02		

Pictorial/Symbolic Decimal Addition 3

☐ = 1   ■ = -1   ▯ = 0.1   ▮ = -0.1   ▫ = 0.01   ▪ = -0.01

The number in the pictorial column plus the number in the decimal column should equal the number in the word column.

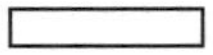

Pictorial Representation +	Decimal Representation	=	Word Representation
■ ☐ ☐ (with small ▪ ▪, ▫ ▫ ▪, ▮ ▮)			Two and forty-six hundredths
	-0.47		Two and three hundredths
☐ ▮ ▮ (with ▪ ▫, ▪ ▪ ▫, ▫ ▫)	-3.46		
	2.17		Negative two and one tenth
■ ▯			Four and seventy-six hundredths

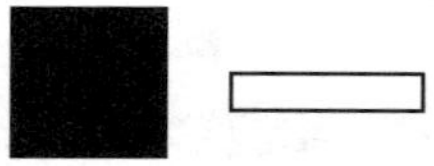

Pictorial/Symbolic Decimal Multiplication 1

☐ = 1   ■ = -1   ☐ = 0.1   ■ = -0.1

The number in the pictorial column time the number in the decimal column should equal the number in the word column.

Pictorial Representation ×	Decimal Representation	Picture and Word Representation =
		Negative twelve hundredths
	0.1	
	-10	
	100	
	-0.01	
	-10	

Pictorial/Symbolic Decimal Multiplication 2

☐ = 1 ■ = -1 ▯ = 0.1 ▮ = -0.1 ▫ = 0.01 ▪ = -0.01

The number in the pictorial column times the number in the decimal column should equal the number in the word column.

Pictorial Representation x	Decimal Representation	= Word Representation
 (3 large black squares, 2 white bars, 3 small black + 1 small white squares)		Twenty-eight and four tenths
	-0.1	Negative four and two tenths
(2 large black squares, 2 black bars, 2 small white + 2 small black squares)	-10	
	0.1	Negative thirty-two hundredths
(1 small black square)		One hundredth

123

Pictorial/Symbolic Decimal Multiplication 3

☐ = 1   ■ = -1   ▭ = 0.1   ▬ = -0.1   □ = 0.01   ■ = -0.01

The number in the pictorial column times the number in the decimal column should equal the number in the word column.

Pictorial Representation x	Decimal Representation	=	Word Representation
			Twenty-eight hundredths
■ ■ □ ■ □ ▬ ▬ ■ ■	- 0.1		
	- 10		Negative twelve

Pictorial Decimal Multiplication/Factoring 1

☐ = 1   ▭ = 0.1   ▫ = 0.01

Symbolic Representation	Pictorial Representation	Factored Pictorial (always forms a rectangle)	Factored Symbolic
1.21	☐ ▭▭ ▫	[rectangle diagram]	(1.1)(1.1) or (1.1)²
1.44		[rectangle diagram]	
1.69			
1.32			
1.96			

Pictorial Decimal Multiplication/Factoring 2

☐ = 1   ▭ = 0.1   ▫ = 0.01

Symbolic Representation	Pictorial Representation	Factored Pictorial (always forms a rectangle)	Factored Symbolic
1.43			$(1+0.3)(1+0.1)$
	 (pictorial shown)		
			$(1+0.1)(1+0.4)$
		 (pictorial shown)	
1.80			

Pictorial Mathematics — Decimals — 126

Pictorial Decimal Multiplication/Factoring 3

☐ = 1   ■ = -1   ▭ = 0.1   ▬ = -0.1   □ = 0.01   ■ = -0.01

Symbolic Representation	Pictorial Representation	Factored Pictorial (always forms a rectangle)	Factored Symbolic
1 - 0.2 + 0.01			(1 - 0.1)(1 - 0.1)
1 - 0.3 + 0.02			
1 - 0.4 + 0.04			
1 - 0.5 + 0.06			
1 + 0.3 - 0.04			

Pictorial Decimal Multiplication/Factoring 4

☐ = 1   ■ = -1   ▭ = 0.1   ▬ = -0.1   ▫ = 0.01   ▪ = -0.01

Symbolic Representation	Pictorial Representation	Factored Pictorial (always forms a rectangle)	Factored Symbolic
1.26			$(1 + 0.4)(1 - 0.1)$
			$(1 - 0.2)(1 - 0.1)$
1.04			

Pictorial Decimal Multiplication/Factoring 5

□ = 1   ■ = -1   ▬ = 0.1   ▭ = -0.1   ▫ = 0.01   ▪ = -0.01

Symbolic Representation	Pictorial Representation	Factored Pictorial (always forms a rectangle)	Factored Symbolic
2.09	□□ ▬▬▬ ▪		
		▬▬▬▬▬ ▪▪ ▬▬▬▬▬ ▪▪ □ □	$(2 - 0.2)(1 + 0.1)$
		▬▬ ▪ □ □	$(2 + 0.3)(1 - 0.1)$

Pictorial Mathematics — Decimals — 129

Pictorial Mathematics   Decimals

Algebraic Decimal Representations 1

☐ = 1   ■ = -1   ▭ = 0.1   ▫ = 0.01
                            ▬ = -0.1   ▪ = -0.01

Shade in the appropriate number of boxes so that when all shaded and un-shaded boxes are added they represent the given expression.

Example: -3.24

1.62

-2.45

4.03

0.02

-1.40

130

Algebraic Decimal Representations 2

☐ = 1   ■ = −1   ▭ = 0.1   ▫ = 0.01
▬ = −0.1   ▪ = −0.01

Shade in the appropriate number of boxes so that when all shaded and un-shaded boxes are added they represent the given expression.

−2.2

4.64

−1.57

3.15

−5.12

−0.02

Pictorial Mathematics — Decimals

Algebraic Decimal Representations 3

☐ = 1   ■ = -1   ▭ = 0.1   ▫ = 0.01
                 ▬ = -0.1   ▪ = -0.01

Shade in the appropriate number of boxes so that when all shaded and un-shaded boxes are added they represent the given expression.

-1.34

3.78

-0.01

4.25

-0.26

5.54

132

1/2	1/2

1/3	1/3	1/3

1/4	1/4	1/4	1/4

1/5	1/5	1/5	1/5	1/5

Fractions

1/6	1/6	1/6	1/6	1/6	1/6

1/8	1/8	1/8	1/8	1/8	1/8	1/8	1/8

1/9	1/9	1/9	1/9	1/9	1/9	1/9	1/9	1/9

1/10	1/10	1/10	1/10	1/10	1/10	1/10	1/10	1/10	1/10

Fractions

What Will Students Be Conceptualizing/Practicing?

- Developing conceptual understanding of a fractional part of a whole

- Developing conceptual understanding of a fractional part of a composite unit

- Learning the model-generating language of fractions

- Identifying/constructing equivalent fractions

- Unitizing: conceptualizing the unit

- Conceptualizing $\frac{1}{2}$: multiple representations of $\frac{1}{2}$

- Conceptualizing $\frac{1}{3}$: multiple representations of $\frac{1}{3}$

- Conceptualizing $\frac{1}{4}$: multiple representations of $\frac{1}{4}$

- Conceptualizing taking a fractional part of a quantity: what is $\frac{a}{b}$ of c? In other words, what is (a ÷ b)(c)

- The meaning of a fraction across several different types of units

- Proportional reasoning: the relative size/value of a unit

- Division of fractions

- Using whole number approximations of fractions to compute multiplications and divisions of fractions

- Using whole number approximations of fractions to compute additions and subtractions of fractions

- Addition of fractions with common denominators

- Addition of fractions with uncommon denominators

- Translating a pictorial multiplication of a proper fraction by a whole number into a numeric representation

- Translating a pictorial multiplication of a mixed fraction by a whole number into a numeric representation

- Transforming a pictorial multiplication of a proper fraction by a whole number

- Translating a pictorial multiplication of a mixed fraction by a mixed fraction into a numeric representation

- Transforming a pictorial multiplication of a mixed fraction by a mixed fraction into a symbolic/numeric representation

- Division of mixed fractions by proper fractions

- Division of mixed fractions by whole numbers

- Division of mixed fractions by mixed fractions

- Percents as a specialized form of a fraction

- Proportions: Using $\frac{1}{100}$ as a referent

- Percent-fraction connections

- Decimal-fraction connections

- Ratio-fraction connections.

Pictorial Templates* Most Useful for the Teaching and Learning of Fractions

- 10 by 10 grids (pp. 345-47)

- 1-inch graph paper (p. 349)

- 4 by 3 rectangular dot paper (p. 366)

- Circle fractions 1, 2, and 4 (pp. 350-352)

- 24 hr clock (p. 353)

- 60-second clock (p. 372)

- Half of half grid (p. 358)

- 4 by 4 square dot paper (p. 359)

- Half-inch graph paper (p. 360)

- Quarter-inch graph paper (p. 361)

- Pictorial worksheets (p. 365)

- Dot picture paper (p. 374)

- 4 by 4 grid (p. 367)

- 3 by 2 grid (p. 368)

*See the back of the appendix for examples on how to use these templates

Teacher Notes
Fractions

"Fractions are infernal beasts created simply to torture children"

Few elementary mathematic concepts present as much difficulty to students and adults as fractions. There are many reasons why fractions prove to be such a difficult web of concepts to master, among them:

1) Fraction concepts and skills are often taught as a set of arbitrary rules that carry little meaning for children. Take division of mixed fractions for example:

   $4 \frac{1}{2} \div 1 \frac{1}{2}$

   Children are taught to:

   - First change $4 \frac{1}{2}$ to an improper fraction
   - Change $1 \frac{1}{2}$ to an improper fraction
   - Invert the second fraction
   - Multiply the numerators
   - Multiply the denominators
   - Reduce the resulting fraction
   - If the result is an improper fraction, change it to a mixed fraction

No wonder so many students are confused by fractions! Why do we change mixed fractions to improper fractions and then change improper fractions back to mixed fractions? Why do we invert the second fraction? Some of these steps are made of five or six other steps. Is there any wonder that students get mixed up, use improper rules, and are simply reduced to thinking they are just not good at math?

2) Fractions are associated with seemingly meaningless terms: numerator, denominator, reducing, improper, proper, mixed fractions. When you combine the fact that fractions are taught as a set of arbitrary rules with the abstract and seemingly terms used in fractions, it is easy to see why the difficulties children encounter with fractions often follow them into adulthood.

Some suggestions to help children develop fraction understanding

1) Spend more time developing meaninful ways of describing fractions before moving on to adding, subtracting, multiplying or dividing fractions. For example, a simple fraction like ¼ should be read in a variety of ways with students:

   a) One out four equal parts
   b) One divided into four equal parts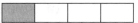
   c) One part out of four equal parts

   d) Selecting one for every four available

Interpreting ¼ as "selecting one for every four" is helpful when students are asked to shade ¼ of a unit that has been divided into eights or sixteenths, etc. For example, the unit below is divided into eights; when students are asked to shade ¼ of the unit, they can interpret ¼ as shading one part of the first row, and one part of the second row.

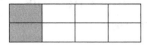

Connecting ¼ with "one out of every four" is also helpful when students are working with parts of a group:

Shade ¼ of the figures

2) Spend more time focusing on problems that get students to think about the concept of a unit. Some of the exercises included in this section that focus students on thinking about the concept of a unit include:

If 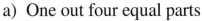 is $\frac{1}{3}$ of a unit, draw the unit

Pictorial Mathematics   Fractions

If ☐☐☐☐☐ is $1\frac{1}{2}$, draw the unit.

If ☐☐☐☐ is $2\frac{2}{3}$ if a unit, draw the unit.

3) Use a variety of pictorial representations to work with fractions. There are several templates included in the appendix that can be used to help students develop fraction understanding. The last part of the appendix also includes sample exercises to use with the templates.

4) Teach your students how to play any one of the four fraction card games included in this section.

5) Take a look at the multiple representation worksheets for the fractions $\frac{1}{2}$, $\frac{1}{3}$ and $\frac{1}{4}$ included in this section. Using these as examples, ask students to create written and pictorial representations for other frequently used fractions such as $\frac{2}{3}, \frac{3}{4}, \frac{1}{5}$ and $\frac{1}{10}$.

6) Use the examples from this section along with the templates in the appendix to create exercises that engage students in transforming and translating fractions across and within each of the five representational systems: Experienced-based, concrete models, pictures, spoken languages, and written symbols.

Developing Fraction Understanding (1)

Picture	Describe what is shaded	Fraction
	1 out of 8 equal parts are shaded	

Pictorial Mathematics   Fractions

Developing Fraction Understanding (2)

Picture	Describe what is shaded	Fraction
	2 out of 8 equal parts are shaded → or, 1 out of four equal parts is shaded →	| |
		| |
		| |
		| |
		| |

Developing Fraction Understanding 3

Picture	Describe what is shaded	Fraction
● ○ ○ ○	1 out of four circles are shaded	
□ ■ ■ ■ ■ ■ ■		
■ ■ ■ ● ○		
△ ○ ● ▲ △		
△ ○ ■ ▲ ○ ○ ▲ ▲		

Developing Fraction Understanding (M)

Picture	Describe what is shaded	Fraction

Pictorial Mathematics   Fractions

Multiple Representations of $\frac{1}{2}$

Put a check (√) above all the items that you think show $\frac{1}{2}$

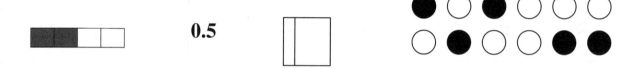

0.5

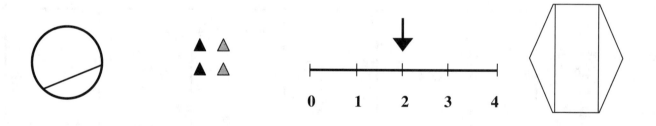

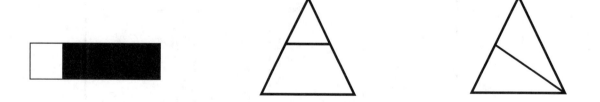

Half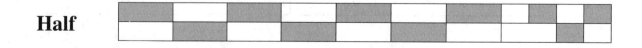

Pictorial Mathematics · Fractions

Multiple Representations of $\frac{1}{3}$

Place a check (√) above all the items that you think show $\frac{1}{3}$

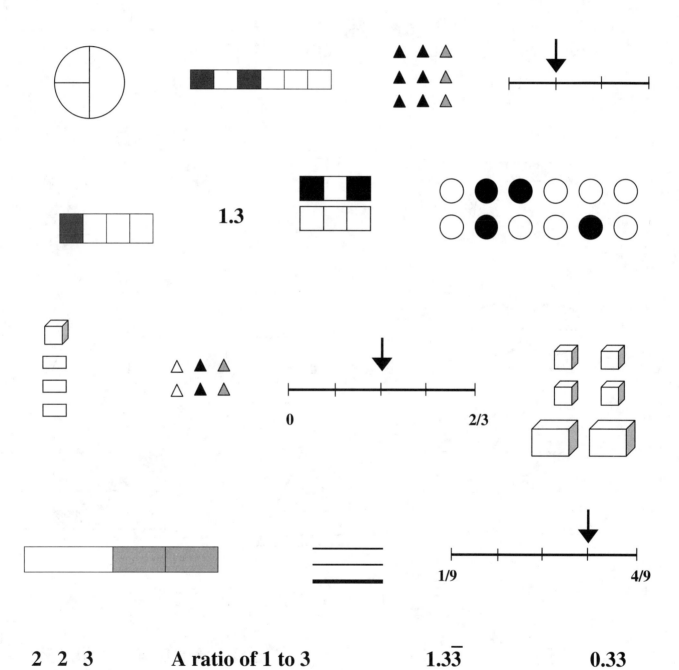

Pictorial Mathematics

Fractions

Multiple Representations of $\frac{1}{4}$

Place a check (√) above all the items that you think show $\frac{1}{4}$

 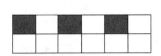 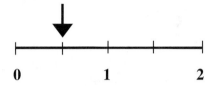

　　0.4　　　　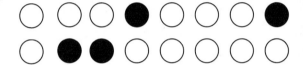

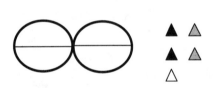 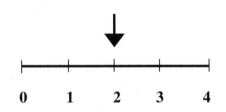 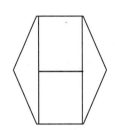

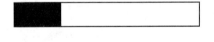  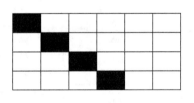

25¢　　　　1.4　　　　A ratio of 1 to 4　　　0.250

146

Pictorial Mathematics — Fractions

Fraction Match (1)

Each figure has parts that are shaded. Indicate which figures have the same fraction of squares shaded, then shade-in the same fraction of squares in figure 5.

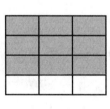

147

Pictorial Mathematics   Fractions

Fraction Match (2)

Each figure has parts that are shaded. Indicate which figures have the same fraction of squares shaded, then shade-in the same fraction of squares in figure 5

  1   2   3   4   5

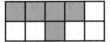

  1   2   3   4   5

  1   2   3   4   5

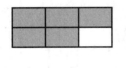   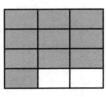   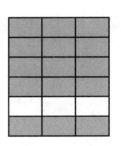

  1   2   3   4   5

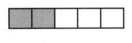 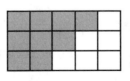   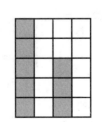

148

Missing Numerators

Circle the number that corresponds to the numerator of the fraction of the shapes that are shaded.

1. ☐☐☐■■   5   3   2   1

2. ☐☐☐■■■   4   6   2   1

3. ●●●●●●○○   2   3   8   4

4. ●●●●●○○○
   ●●●●●○○○   6   8   16   5

5. ●●○○○○○○
   ●●○○○○○○   16   1   12   3

6. ▲△▲▲△▲
   ▲△▲▲△▲   2   12   3   4

Pictorial Mathematics — Fractions

Missing Denominators

Circle the number that corresponds to the denominator of the fraction of the shapes that are shaded.

1.    5   3   2   1

2. 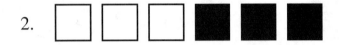   4   8   2   1

3. ●●●●●●○○○   2   3   8   6

4. ●●●●●●●○
   ●●●●●●●○   7   2   14   8

5. ●●○○●●●●
   ●●○○●●●●   12   10   4   3

6. ▲△▲△△△
   ▲△▲△△△   4   8   3   2

Developing Fraction Understanding
Changing Units (1)

Shade-in $\frac{1}{4}$ of each grid

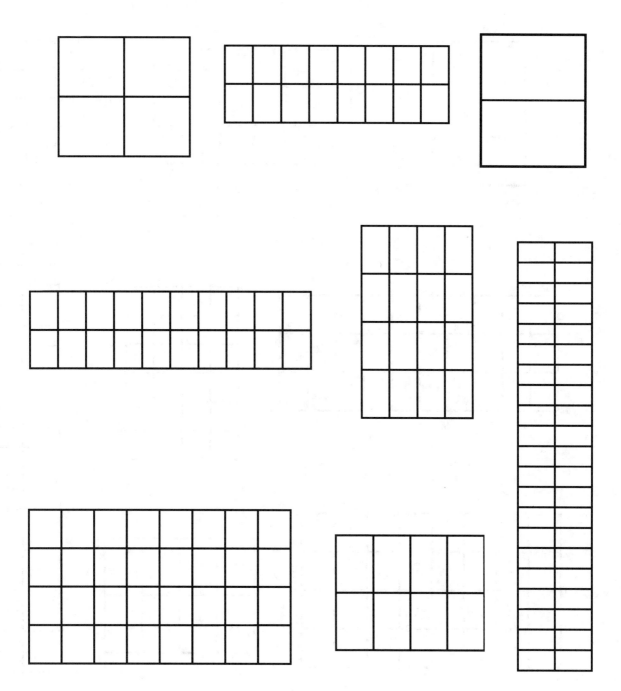

Developing Fraction Understanding
Changing Units (2)

Shade-in $\frac{2}{3}$ of each grid

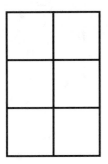
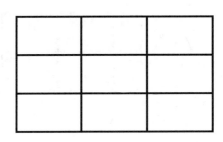
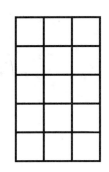

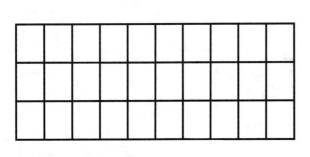
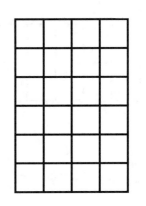

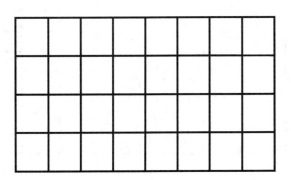
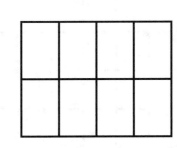

Pictorial Mathematics — Fractions

Developing Fraction Understanding
Changing Units (3)

Shade-in the fraction given for each grid

$\dfrac{5}{8}$

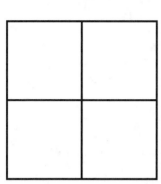

$\dfrac{1}{2}+\dfrac{1}{4}$

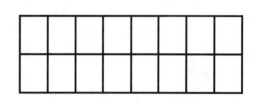

$\dfrac{3}{5}$

$\dfrac{1}{8}+\dfrac{1}{4}$

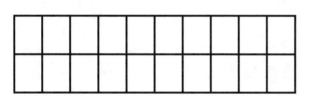

$\dfrac{1}{16}+\dfrac{1}{8}$

$\dfrac{5}{16}$

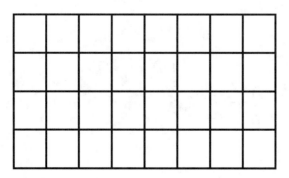
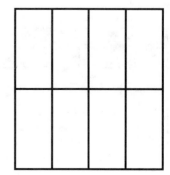

153

Pictorial Mathematics   Fractions

Estimating Fractions 1

Circle the fractions that most closely represent the shaded part of the figure.

   a) $\dfrac{3}{8}$ b) $\dfrac{1}{5}$ c) $\dfrac{1}{10}$ d) $\dfrac{1}{6}$

    a) $\dfrac{1}{3}$ b) $\dfrac{1}{4}$ c) $\dfrac{2}{5}$ d) $\dfrac{3}{16}$

    a) $\dfrac{1}{5}$ b) $\dfrac{2}{7}$ c) $\dfrac{2}{3}$ d) $\dfrac{2}{5}$

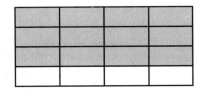   a) $\dfrac{7}{8}$ b) $\dfrac{11}{15}$ c) $\dfrac{9}{10}$ d) $\dfrac{17}{20}$

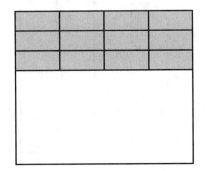   a) $\dfrac{12}{37}$ b) $\dfrac{4}{15}$ c) $\dfrac{2}{9}$ d) $\dfrac{3}{7}$

Pictorial Mathematics   Fractions

Estimating Fractions 2

Circle the fractions that most closely represent the shaded part of the figure.

   a) $\frac{3}{5}$ b) $\frac{2}{3}$ c) $\frac{4}{5}$ d) $\frac{7}{8}$

   a) $\frac{7}{8}$ b) $\frac{2}{4}$ c) $\frac{4}{5}$ d) $\frac{3}{5}$

   a) $\frac{1}{3}$ b) $\frac{1}{8}$ c) $\frac{1}{5}$ d) $\frac{2}{8}$

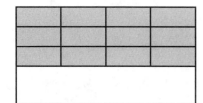   a) $\frac{9}{14}$ b) $\frac{3}{4}$ c) $\frac{1}{2}$ d) $\frac{5}{12}$

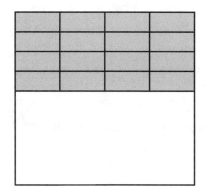   a) $\frac{12}{29}$ b) $\frac{5}{11}$ c) $\frac{8}{17}$ d) $\frac{6}{13}$

Pictorial Mathematics   Fractions

From Fractions to Units (1)

If ▪▪/▪▪ is $\frac{2}{3}$ of the unit, draw the unit

If ▪▪▪▪▪ is $\frac{5}{8}$ of the unit, draw the unit

If ▪▪▪▪▪▪ is $1\frac{1}{3}$ of a unit, draw the unit

If ▲▲▲ ▲▲▲ is $\frac{3}{4}$ of a unit, draw the unit

If |—|—|—| is $1\frac{1}{5}$ of a unit, draw the unit

If ○○○ is the same as 4 units, draw the unit.

156

Pictorial Mathematics   Fractions

From Fractions to Units (2)

1. If ▪▪/▪▪ is $1\frac{1}{3}$ of a unit, draw the unit

2. If ▪▪▪▪▪ is $1\frac{1}{4}$ of a unit, draw the unit

3. If ▪▪▪▪▪▪ is $2\frac{2}{5}$ of a unit, draw the unit

4. If ▲▲▲▲ / ▲▲▲▲ is $1\frac{1}{3}$ of a unit, draw the unit

5. If |—|—|—|—|—|—| is $2\frac{1}{5}$ of a unit, draw $1\frac{3}{5}$ of a unit

6. If ○○○ / ○○○ / ○○○ / ○○○ is the same as 8 units, draw 5 units

157

From Fractions to Units (3)

1. If ▪▪▪▪▪▪▪▪ is $2\frac{2}{3}$ of the unit, draw $\frac{1}{3}$ of the unit

2. If ▪▪▪▪▪▪▪▪▪▪▪▪▪ is $3\frac{3}{4}$ of the unit, draw $\frac{1}{2}$ of the unit

3. If ▪▪▪▪▪▪▪▪▪▪▪ is $1\frac{5}{6}$ of a unit, draw $\frac{1}{3}$ of the unit

4. If ▲▲▲▲▲ ▲▲▲▲▲ is $2\frac{1}{2}$ of a unit, draw the unit

5. If 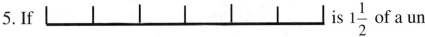 is $1\frac{1}{2}$ of a unit, draw $1\frac{1}{4}$ of a unit

6. If 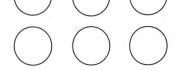 is the same as $6\frac{2}{3}$ units, draw $1\frac{1}{3}$ units

Pictorial Mathematics   Fractions

Fraction Partitioning (1)

1. Four children want to share these five chocolate bars equally. Shade-in one child's part.

   ☐ ☐ ☐ ☐ ☐

2. What fraction of one chocolate bar does each child get? _____

3. Eight children share these three cookies equally. Shade-in one child's part.

   ☐ ☐ ☐

4. What fraction of one cookie does each child get? _____

5. Four children share these ten candy bars equally. Shade-in one child's part.

   ☐ ☐ ☐ ☐ ☐
   ☐ ☐ ☐ ☐ ☐

6. What fraction of one candy bar does each child get?

7. Shade-in $\frac{5}{8}$ of the candy bar below.

Fraction Partitioning (2)

1. Six children want to share these four candy bars equally. Shade-in one child's part.

2. What fraction of one candy bar does each child get? _____

3. Four children want to share these three cookies equally. Shade-in one child's part.

4. What fraction of one cookie does each child get? _____

5. Nine children want to share these twelve candy bars equally. Shade-in one child's part.

6. What fraction of one candy bar does each child get? _____

7. Use pictures to show how many $1\frac{3}{4}$ bars of candy can be cut from 12 bars.

Pictorial Mathematics   Fractions

Fraction Size Sense (overhead 1)

Students are going to select one of six possible answers by holding the appropriate number of fingers next to their chest (0 fingers means I don't know, 1 finger means the answer is a, 2 fingers b, 3 is c, 4 is d, and 5 is e). Students should explain their reasoning.

Using only mental estimation, which point is closest to the approximate answer?

$2\frac{1}{10}$ divided by $\frac{37}{9}$

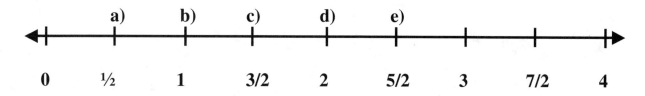

$3\frac{7}{8}$ times $\frac{2}{9}$

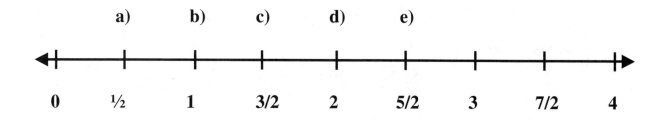

$\frac{1}{100}$ divided by 50

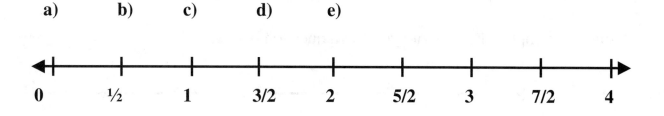

Pictorial Mathematics · Fractions

Fraction Size Sense (overhead 2)

Using only mental estimation, which point is closest to the approximate answer?

$2\frac{2}{30}$ divided by $\frac{3}{13}$

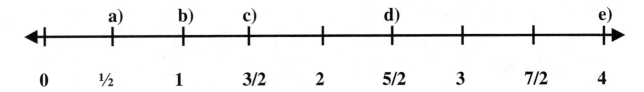

$2\frac{17}{18}$ divided by $\frac{20}{7}$

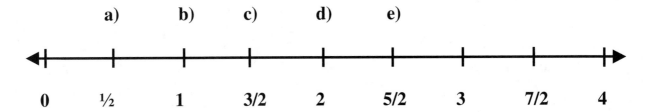

$\frac{41}{50}$ times 50 divided by 19

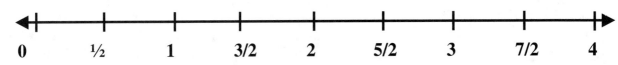

Using the samples above, write your own question below

162

Pictorial Mathematics Fractions

Fraction Size Sense (overhead 3)

Using only mental estimation, which point is closest to the approximate answer?

$2\frac{1}{10} + \frac{8}{9}$

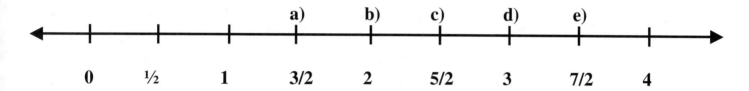

$3\frac{7}{8} - \frac{21}{11}$

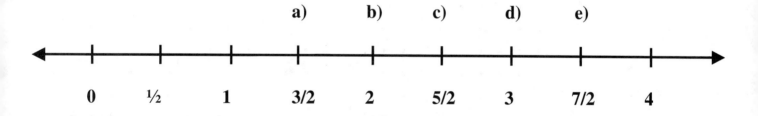

$\frac{21}{10} \times \frac{9}{5}$

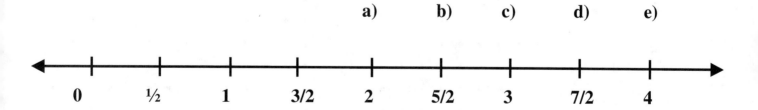

Adding Fractions. Common Denominators (1)

Use pictures to do each of the exercises on the left side. Write the answer in fraction form next to your pictures. The number in parenthesis tells you how many small squares should be in your unit.

¼ + ¼
(4)

$\frac{1}{3} + \frac{1}{3}$
(3)

$\frac{1}{6} + \frac{2}{6}$
(6)

$\frac{1}{8} + \frac{3}{8}$
(8)

$\frac{2}{9} + \frac{5}{9}$
(9)

$\frac{3}{10} + \frac{1}{10}$
(10)

Adding Fractions. Common Denominators (2)

Use pictures to do each of the exercises on the left side. Write the answer in fraction form next to your pictures. The number in parenthesis tells you how many small squares should be in your unit.

¼ + ¾
(4)

$\dfrac{1}{3} + \dfrac{4}{3}$
(3)

$\dfrac{5}{6} + \dfrac{5}{6}$
(6)

$\dfrac{7}{8} + \dfrac{3}{8}$
(8)

$\dfrac{13}{9} + \dfrac{7}{9}$
(9)

Pictorial Mathematics — Fractions

Adding Fractions. Common Denominators (3)

Use pictures to do each of the exercises on the left side. Write the answer in fraction form next to your pictures.

$\dfrac{5}{6} + \dfrac{7}{6}$

$\dfrac{1}{3} + \dfrac{4}{3} + \dfrac{5}{3}$

$\dfrac{7}{6} + \dfrac{5}{6} + \dfrac{5}{6}$

$\dfrac{13}{8} + \dfrac{3}{8} + \dfrac{5}{8}$

$\dfrac{11}{10} + \dfrac{7}{10} + \dfrac{3}{10}$

Pictorial Mathematics   Fractions

Adding Fractions (1)

Use pictures to do each of the exercises on the left side. Write the answer in fraction form next to your pictures. The number in parenthesis tells you how many small squares should be in your unit.

$\dfrac{3}{4} + \dfrac{1}{2}$
(4)

$1\dfrac{1}{2} + 1\dfrac{1}{4}$
(4)

$\dfrac{2}{3} + \dfrac{5}{6}$
(6)

$\dfrac{7}{8} + \dfrac{1}{2}$
(8)

$\dfrac{5}{9} + \dfrac{2}{3}$
(9)

Adding Fractions (2)

Use pictures to do each of the exercises on the left side. Write the answer in fraction form next to your pictures.

$1\dfrac{3}{4} + \dfrac{4}{3}$

$\dfrac{3}{8} + \dfrac{5}{6}$

$\dfrac{7}{12} + \dfrac{5}{8}$

$\dfrac{4}{3} + \dfrac{7}{6}$

$\dfrac{11}{9} + \dfrac{11}{6}$

$\dfrac{7}{5} + \dfrac{2}{3}$

Adding Fractions (3)

Use pictures to do each of the exercises on the left side. Write the answer in fraction form next to your pictures.

$1\dfrac{3}{4} + 1\dfrac{1}{2}$

$\dfrac{7}{8} + \dfrac{5}{4}$

$\dfrac{4}{3} + \dfrac{7}{6}$

$\dfrac{10}{9} + \dfrac{11}{6}$

$\dfrac{7}{9} + \dfrac{2}{3}$

Pictorial Mathematics   Fractions

Adding Fractions (4)

Use pictures to do each of the exercises on the left side. Write the answer in fraction form next to your pictures.

$2\dfrac{1}{3} + 1\dfrac{1}{2}$

$\dfrac{5}{8} + \dfrac{5}{6}$

$\dfrac{7}{8} + \dfrac{5}{12}$

$\dfrac{7}{9} + \dfrac{5}{6}$

$1\dfrac{4}{9} + 1\dfrac{5}{6}$

Adding Fractions (5)

Use pictures to do each of the exercises on the left side. Write the answer in fraction form next to your pictures.

$\dfrac{1}{2} + \dfrac{3}{5} + \dfrac{7}{10}$

$\dfrac{3}{8} + \dfrac{1}{4} + \dfrac{1}{2}$

$\dfrac{1}{3} + \dfrac{3}{5} + \dfrac{4}{15}$

$\dfrac{7}{9} + \dfrac{2}{3} + \dfrac{1}{6}$

Subtracting Fractions, Common Denominators (1)

Use pictures to do each of the exercises on the left side. Write the answer in fraction form next to your pictures.

$\dfrac{3}{4} - \dfrac{1}{4}$

$\dfrac{7}{9} - \dfrac{4}{9}$

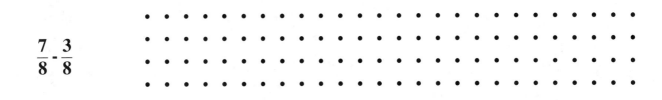

$\dfrac{7}{8} - \dfrac{3}{8}$

$\dfrac{7}{10} - \dfrac{3}{10}$

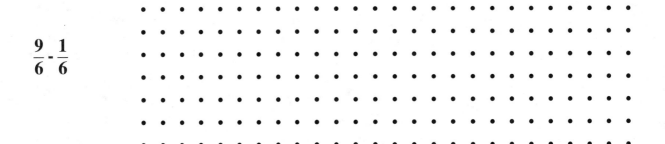

$\dfrac{9}{6} - \dfrac{1}{6}$

Pictorial Mathematics   Fractions

Subtracting Fractions, Common Denominators (2)

Use pictures to do each of the exercises on the left side. Write the answer in fraction form next to your pictures.

$3\frac{1}{6} - 1\frac{5}{6}$

Pictorial Mathematics   Fractions

Subtracting Fractions, Uncommon Denominators (1)

Use pictures to do each of the exercises on the left side. Write the answer in fraction form next to your pictures.

$\dfrac{1}{3} - \dfrac{1}{6}$ I drew one unit made of 6 rectangles. I shaded $\dfrac{1}{3}$ with horizontal lines. Then I shaded $\dfrac{1}{6}$ with vertical lines. The rectangle that ended up with both vertical and horizontal shades is the answer. So, $\dfrac{1}{3} - \dfrac{1}{6} = \dfrac{1}{6}$

$\dfrac{2}{3} - \dfrac{2}{9}$

$\dfrac{3}{4} - \dfrac{3}{8}$

$\dfrac{5}{8} - \dfrac{5}{6}$

$\dfrac{7}{9} - \dfrac{1}{3}$

174

Pictorial Mathematics   Fractions

Subtracting Fractions, Uncommon Denominators (2)

Use pictures to do each of the exercises on the left side. Write the answer in fraction form next to your pictures.

 $1\frac{1}{3} - \frac{5}{6}$

 $3\frac{1}{4} - 1\frac{2}{3}$

 $3\frac{3}{5} - 1\frac{1}{2}$

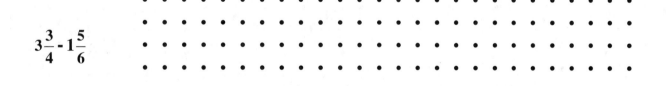 $3\frac{3}{4} - 1\frac{5}{6}$

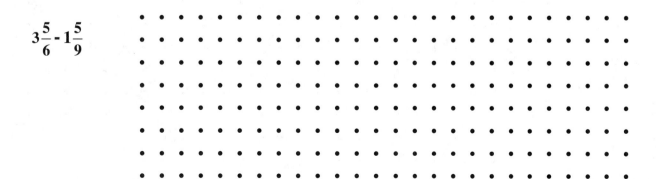 $3\frac{5}{6} - 1\frac{5}{9}$

Subtracting Fractions, Uncommon Denominators (3)

Use pictures to do each of the exercises on the left side. Write the answer in fraction form next to your pictures.

$2\frac{1}{6} - \frac{2}{3}$

$2\frac{1}{3} - 1\frac{2}{3}$

$2\frac{2}{3} - 1\frac{3}{4}$

$3\frac{3}{4} - 2\frac{7}{8}$

$2\frac{5}{8} - 1\frac{5}{6}$

Pictorial Mathematics | Fractions

Subtracting Fractions, Uncommon Denominators (4)

Use pictures to fo each of the exercises on the left side. Write the answer in fraction form next to your pictures.

$1\dfrac{1}{3} - \dfrac{5}{6}$

$3\dfrac{1}{4} - 1\dfrac{2}{3}$

$3\dfrac{3}{5} - 1\dfrac{1}{2}$

$3\dfrac{3}{4} - 1\dfrac{5}{6}$

$3\dfrac{5}{6} - 1\dfrac{5}{9}$

Repeating Groups (1)

Example:

 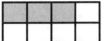 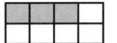　　$3 \times \dfrac{3}{8} = \dfrac{9}{8} = 1\dfrac{1}{8}$

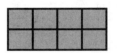 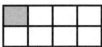

1)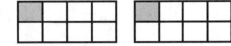

　　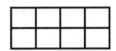

2) 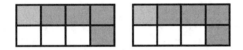　　

3)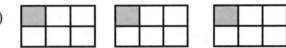

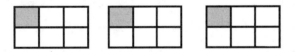　　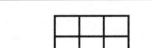

4) 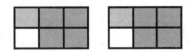　　

5)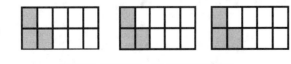

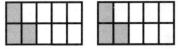　　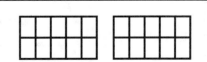

Pictorial Mathematics | Fractions

Repeating Groups (2)

Example: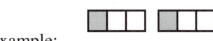

$2 \times \dfrac{1}{3} = \dfrac{2}{3}$

1)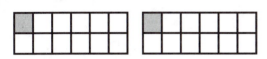

2)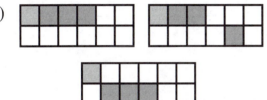

3)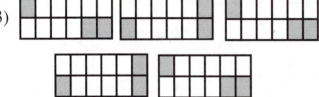

4)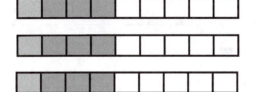

5)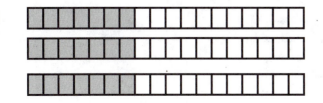

Repeating Groups (3)

Example:

$$2 \times \frac{2}{3} = \frac{4}{3} = 1\frac{1}{3}$$

1)

2)

3)

4)

5)

Multiplication of Mixed Fractions 1

Example:

$2 \times 1\dfrac{2}{3} = 2 \times \dfrac{5}{3} = \dfrac{10}{3} = 3\dfrac{1}{3}$

1) _____

2) _____

3) _____

4) _____

5) _____

Multiplication of Mixed Fractions (2)

Example:

$$2 \times 1\frac{3}{4} = 2 \times \frac{7}{4} = \frac{14}{4} = 3\frac{2}{4} = 3\frac{1}{2}$$

1)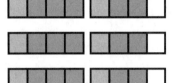

2)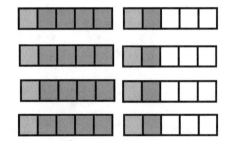

3)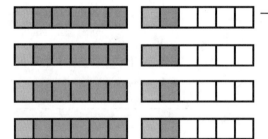

4)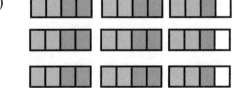

Multiplication of Mixed Fractions (3)

Example:

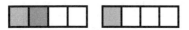

$1\frac{1}{2} \times 1\frac{1}{2}$ 1 and a half repeated one and a half times.

$1\frac{1}{2}$ groups of $1\frac{1}{2}$ units make a total of $2\frac{1}{4}$ units.

**What multiplication is being shown?
What is the total?**

1)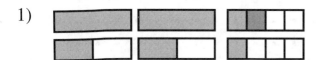

2)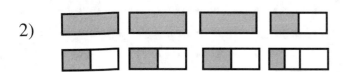

3)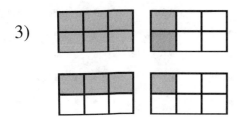

4)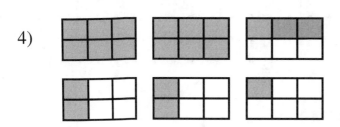

Pictorial Mathematics — Fractions

Multiplying Fractions (1)

Example

Use pictures to show $4 \times 1\frac{1}{2}$

(4 groups of $1\frac{1}{2}$)

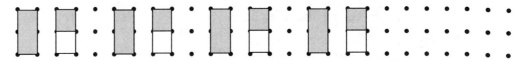

$4 \times 1\frac{1}{2}$ is the same as four groups of $1\frac{1}{2}$ or $1\frac{1}{2}$ repeated 4 times, or 6 wholes

Use pictures to show $1\frac{1}{2} \times 4$)

($1\frac{1}{2}$ groups of 4)

Use pictures to show $\frac{1}{2} \times 2\frac{1}{2}$

(half of $2\frac{1}{2}$)

Use pictures to show $1\frac{1}{2} \times \frac{1}{4}$

($1\frac{1}{2}$ groups of $\frac{1}{4}$)

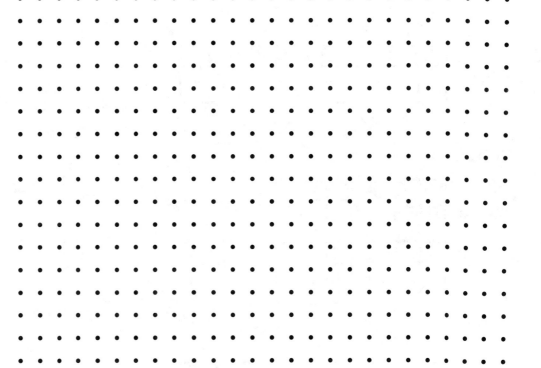

Multiplying Fractions (2)

Use pictures to show

$3\frac{1}{2} \times 1\frac{1}{2}$

($3\frac{1}{2}$ groups of $1\frac{1}{2}$)

Use pictures to show

$1\frac{1}{2} \times 3\frac{1}{2}$

($1\frac{1}{2}$ groups of $3\frac{1}{2}$)

Use pictures to show

$2\frac{1}{2} \times 3\frac{3}{4}$

($2\frac{1}{2}$ groups of $\frac{4}{3}$)

Pictorial Mathematics Fractions

Multiplying Fractions (3)

Use pictures to show

$3\frac{1}{2} \times 2\frac{1}{2}$

(____ groups of ____)

Use pictures to show

$2\frac{1}{2} \times \frac{5}{3}$

(____ groups of ____)

Use pictures to show

$3\frac{1}{2} \times 2\frac{3}{4}$

(____ groups of ____)

186

Teacher Notes – Division of Fractions

Just like in multiplication, the language used by the teacher plays a decisive role in the type of conceptual understanding students develop about multiplication. To illustrate, let's look at two elementary fraction divisions (1 and 2) and one more conceptually complex division of fractions (3):

Example 1) $3 \div \frac{1}{2}$

This division can and should be read in a variety of different ways by both teachers and students:

- What do you get when you divide 3 wholes into $\frac{1}{2}$ pieces?
- How many half pieces make 3 wholes?
- How many halves add up to 3 wholes?
- How many times can you subtract exactly $\frac{1}{2}$ pieces from 3 wholes?

The teacher should then lead the students through the problem with a pictorial representation, as follows:

This shows 3 wholes

This shows that when you divide 3 wholes into halves, you get a total of 6 halves.

Ask students why they get a larger number (6) than the one they started with (3) when dividing 3 by $\frac{1}{2}$. Aren't we supposed to get smaller numbers when we divide and larger numbers when we multiply? Have students practice dividing by a fraction smaller than one by folding and cutting three equal small sheets of paper into halves. Give students several similar problems, such as two divided by one fourth, one and one third divided by one third, etc.

Example 2) $1\frac{1}{4} \div 5$

In this division, unlike with the first example, we are dividing a small number by a larger number. The best way to read this type of division is:

$$1\frac{1}{4} \div 5$$

When we divide $1\frac{1}{4}$ into five equal parts, how large is each piece?

Again, the teacher should then lead the students through the problem with a pictorial representation.

   This shows $1\frac{1}{4}$ using five $\frac{1}{4}$ pieces within 2 wholes

Teachers should help students realize that the picture shows 5 equal pieces, each of $\frac{1}{4}$ size, totaling $1\frac{1}{4}$. Thus, when $1\frac{1}{4}$ is divided into five equal parts, each part is $\frac{1}{4}$.

Before moving on to the third example, note that the language in the second example is quite different from the language in the first example. Teachers must make an effort to use the proper language for the conceptual model being taught. Using the appropriate conceptual language consistently takes a little practice and effort, but the alternatives are to teach only rules and procedures without much meaning, or to use language that does not scaffold students' understanding.

Example 3) $1\frac{1}{8} \div \frac{3}{4}$

This division is more complex than the two previous ones because $\frac{3}{4}$ does not go evenly into $1\frac{1}{8}$. This division could be read as follows:

- How many $\frac{3}{4}$ pieces make up $1\frac{1}{8}$?
- How many $\frac{3}{4}$ pieces can be made from $1\frac{1}{8}$?

As with all examples, the teacher should then lead the students through the problem with a pictorial representation.

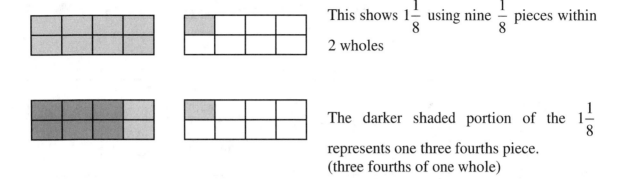

This shows $1\frac{1}{8}$ using nine $\frac{1}{8}$ pieces within 2 wholes

The darker shaded portion of the $1\frac{1}{8}$ represents one three fourths piece. (three fourths of one whole)

Look at the dark pieces that represent the $\frac{3}{4}$. Remind students that they are being asked how many $\frac{3}{4}$ pieces can be taken out from $1\frac{1}{8}$. Students should see that they can take one piece of $\frac{3}{4}$ size from $1\frac{1}{8}$, but that there is not enough to take out two $\frac{3}{4}$ pieces. The question is, what part of $\frac{3}{4}$ are the remaining three light-shaded pieces? Since $\frac{3}{4}$ is made of six shaded pieces, the remaining three light pieces make $\frac{1}{2}$ of $\frac{3}{4}$. Teachers can then lead the students through putting all this information together.

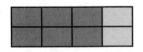 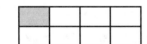

$1\frac{1}{8}$ is made out of one $\frac{3}{4}$ piece and $\frac{1}{2}$ of a $\frac{3}{4}$ piece. So, the answer to the division ($1\frac{1}{8}$ divided by $\frac{3}{4}$) s $1\frac{1}{2}$.

Model-Generating Language

The exercises in the fraction's chapter use a variety of ways to describe fractions. These descriptions, if used often, can help students connect a particular operation with the proper conceptual model. Asking students how many halves are in $2\frac{1}{2}$ carries far more contextual information than simply asking them to compute $2\frac{1}{2} \div \frac{1}{2}$.

I encourage teachers and parents to use the pictorial worksheet templates included in the appendix to create additional division of fraction exercises. I strongly recommend that teachers and parents solve the problems they create before giving any exercises to their students. This will ensure that students only tackle workable exercises.

Division of Fractions (1)

Use pictures to show 2 ½ units divided into groups of ½ units.

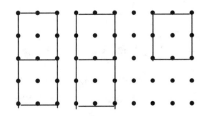

The picture shows $2\frac{1}{2}$ units, divided into $\frac{1}{2}$ unit pieces. $2\frac{1}{2} \div \frac{1}{2} = 5$

Use pictures to show how many $\frac{2}{3}$ would cover 4 units

Show 5 divided by $1\frac{1}{4}$

Show 2 divided by $\frac{1}{5}$

Division of Fractions (2)

Use pictures to show $3\frac{3}{4}$ units divided into groups of $1\frac{1}{2}$ units

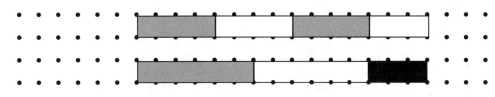

The first picture shows $3\frac{3}{4}$ units. The second picture shows $3\frac{3}{4}$ units divided into $2\frac{1}{2}$ groups of $1\frac{1}{2}$ (check to see that the black shaded piece at the end of the second picture is in fact half of $1\frac{1}{2}$). Therefore,

$3\frac{3}{4} \div 1\frac{1}{2} = 2\frac{1}{2}$

Use pictures to show how many $\frac{2}{3}$ would cover $2\frac{1}{3}$

Show 6 divided by $1\frac{1}{4}$

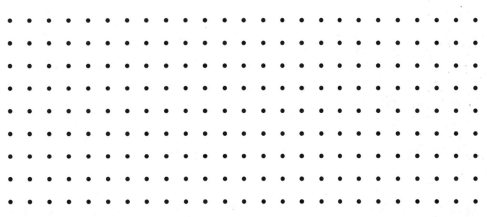

Division of Fractions (3)

Use pictures to show $2\frac{1}{2}$ units divided into groups of $\frac{1}{2}$ units.

Use pictures to show how many $\frac{2}{3}$ would cover 4 units.

Show 5 divided by $1\frac{1}{4}$.

Developing Percent Understanding
Changing Units (1)

For each grid, shade-in the percent of space given

48%   **16%**

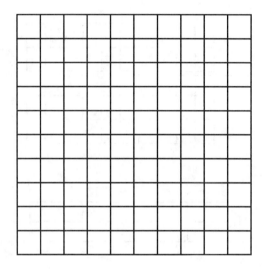

65%

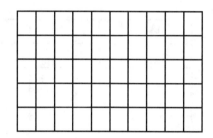

40%

22%   **75%**

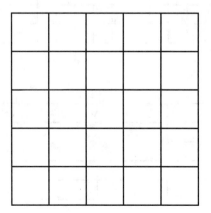
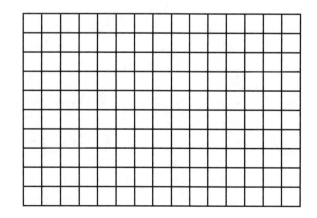

Developing Percent Understanding Changing Units (2)

Shade-in 40% of each grid

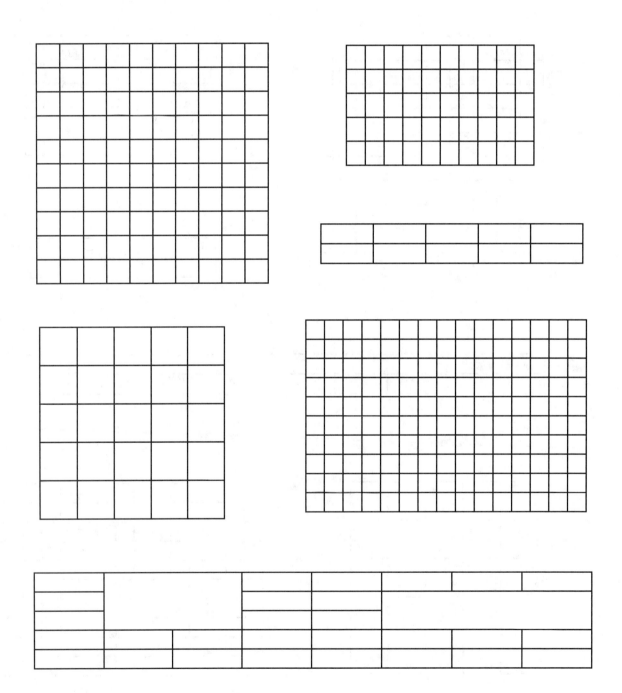

Developing Percent Understanding
Changing Units (3)

For each grid, shade-in the percent of space given

24%   **40%**

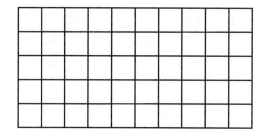

35%   **75%**

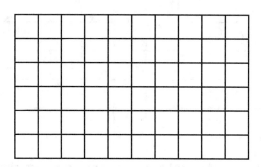

12.5%

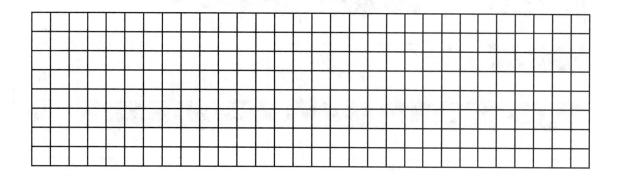

Relation-Strips 1

a) If the length of strip C is 1 unit, how long is A? _____

b) If the length of strip B is $\frac{4}{5}$ of 1 unit, how long is C? _____

c) If the length of strip D is 0.52 of 1 unit, how long is strip B? _____

d) What percent of strip E equals the length of strip D? _____

e) What percent of B would equal twice the length of F? _____

f) If E is 6.5 units, what two strips combined make 4.75 units? _____

g) If A + B + C is one unit, how long is G?

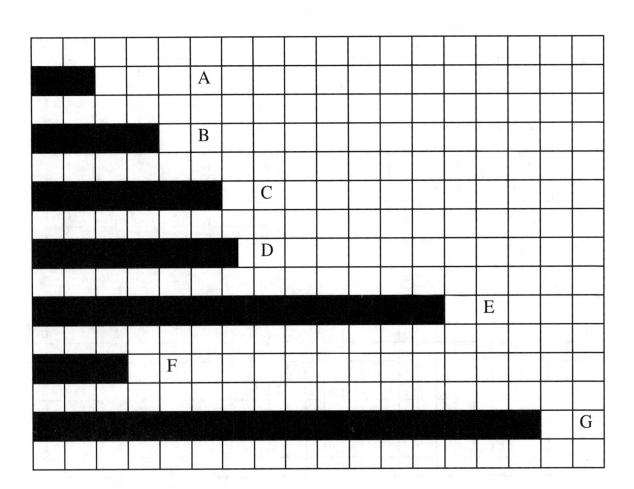

Pictorial Mathematics — Fractions

Relation-Strips 2

h) If the length of strip A is 1 unit, how long is B? _____

i) If the length of strip A is $\frac{1}{2}$ of 1 unit, how long is C? _____

j) If the length of strip D is 0.65 of 1 unit, how long is strip B? _____

k) What percent of strip E equals to $\frac{2}{3}$ of the length of strip A? _____

l) What percent of B would equal twice the length of F? _____

m) If E is $1\frac{3}{4}$ units, what two strips combined make $2\frac{1}{8}$ units? _____

n) B + C - A is one unit, how long is G?

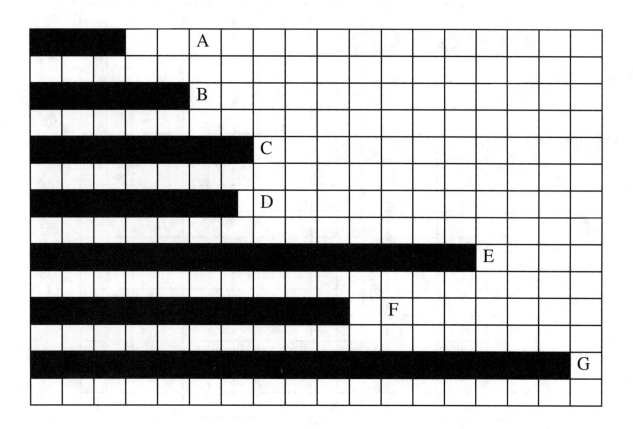

Relation-Strips 3

o) Shade in enough squares so that strip B is $1\frac{1}{4}$ times the size of A

p) Shade in enough squares so that strip C is 50% larger than A

q) Shade in enough squares so that strip D is $1\frac{1}{2}$ times larger than C

r) Shade in enough squares so that strip E is 75% the size of A

s) Shade in enough squares so that strip F is $\frac{1}{5}$ as long as D and E combined.

t) Shade in enough squares so that strip G is 60% as long as D and E combined.

u) Shade in enough squares so that strip H + H is 25% longer than A.

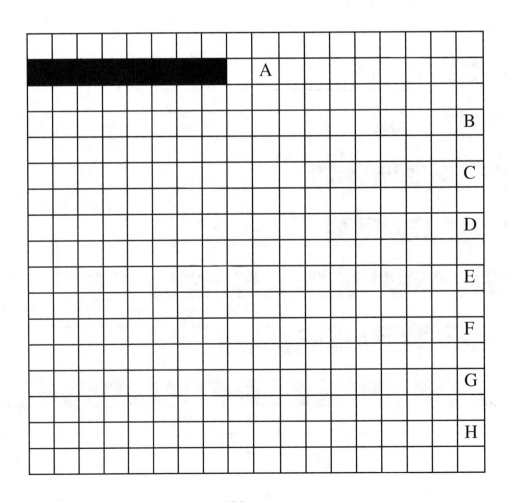

From Percents to Units (1)

1. If ▪▪▪▪▪▪ is 40% of the unit, draw the unit.

2. If ▪▪▪▪▪▪▪▪▪▪▪▪▪▪▪ is 125% of the unit, draw 75% of the unit.

3. If ▪▪▪▪▪▪▪▪▪▪▪▪ is 400% of the unit draw 150% the unit.

4. If ▲▲▲▲▲▲▲▲ is 56% of the unit, draw 21% of the unit.

5. If |⎯|⎯|⎯|⎯|⎯|⎯| is 30% of a unit, draw 20% of a unit.

6. If ○○○○○○○○ is the same as 160% of a unit, draw 5% of the unit.

Pictorial Mathematics | Fractions

From Percents to Units (2)

1. If 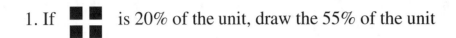 is 20% of the unit, draw the 55% of the unit

2. If 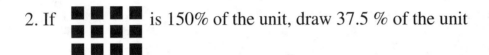 is 150% of the unit, draw 37.5 % of the unit

3. If 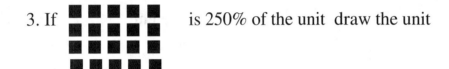 is 250% of the unit draw the unit

4. If 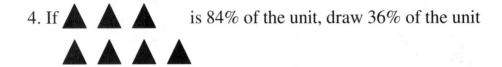 is 84% of the unit, draw 36% of the unit

5. If 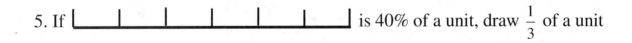 is 40% of a unit, draw $\frac{1}{3}$ of a unit

6. If 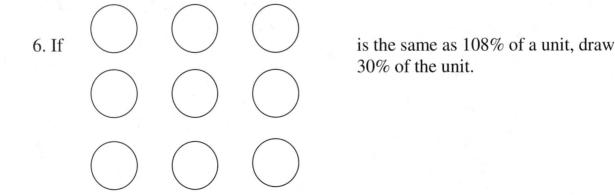 is the same as 108% of a unit, draw 30% of the unit.

Pictorial Mathematics   Fractions

From Percents to Units (3)

1. If ▪▪▪▪▪▪▪ is 48% of the unit, draw 30% of the unit

2. If ▪▪▪▪▪▪▪▪▪▪▪▪▪▪ is 270% of the unit, draw 54% of the unit

3. If ▪▪▪▪▪▪▪▪▪▪▪▪▪▪▪▪▪▪ is 120% of the unit, draw 20% of the unit

4. If ▲▲▲ ▲▲▲▲ is 84% of the unit, draw 36% of the unit.

5. If |⎯|⎯|⎯|⎯|⎯|⎯| is 120% of a unit, draw 80% of a unit.

6. If ○○○ ○○○ ○○○ is the same as 216% of a unit, draw 30% of the unit.

201

Developing Fraction / Percent Connections

2A	B	12	F	C
45	4W	15	A	2A
ABC	FC	5F	68	P
FA	7F	3F	WF	A

1. What fraction of the rectangles have only numbers inside?

2. What percent of the rectangles have both letters and numbers inside?

3. How many rectangles have only letters inside?

4. What fraction of the rectangles that have letters have an A inside?

5. Write a question using the table above such that the answer is 28%.

6. Write a question using the table above such that the answer is $\frac{3}{7}$.

7. Write a question using the table above such that the answer is 95%.

8. Fill-in the table below so that $\frac{2}{5}$ of the rectangles have numbers larger than 20, 45% have only letters and 15% have both numbers and letters.

Fraction Card Games

The following are some examples of the type of learning-games students can play with the fraction cards included in this section to further develop their conceptual understanding.

There are 90 cards, or two sets of 45. One set is made up of numerical fractions (⅓, ¼, etc.), equivalent decimals and percents; the other set has the same fractions, but they are in pictorial form (see below). For each fraction (i.e. ¼) there are eight equivalent values, four pictorial ones and for numeric ones. There is one wild card that can be paired with any card. The teacher should split the 90 cards into two decks, making sure to include all the equivalent cards in each set.

Fraction Poker
Students play in groups of three to five. The first few times the teacher might play with the whole class by copying a set of two player's cards on transparencies. Each player gets a total of five cards. The object of the game is to make the highest pair, three of a kind, or four of a kind possible. For example:

Player 1

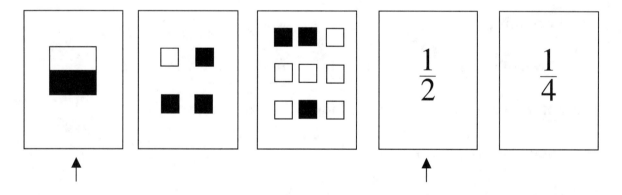

This player has one pair of cards showing the same fraction, ½. He or she will keep this pair, and ask the dealer to change their other three cards to improve their game.

Player 2

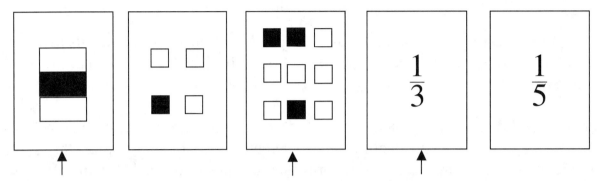

Player 2 has three cards with the same fraction, $\frac{1}{3}$. This player is said to have three of a kind, which always beats any pair. As it stands right now, player 2 has a better hand than player 1. This player will keep the three equivalent cards and exchange the two other cards.

Here is the order of best possible hands, starting with the worse possible hand, not even one pair, and ending with the best possible hand, four of a kind.

1) **No pair.** When nobody has a pair, the player with the highest card wins. The highest card in the deck is 1.

2) **One pair.** (As before, if two players have a pair, the player with highest value pair wins. For example, a pair of halves, beats a pair of thirds.

3) **Two pairs**

4) **Three of a kind**

5) **Full house, or three of kind along with two of a kind**

6) **Poker, or four of a kind**

Poker variations

- Give the players chips to bet. Each player "anties", that is, puts in 1-3 chips before they get their cards. They can bet after they get their first five cards and they can make a final bet after exchanging cards.

- **Hold them**. Players get two cards. They bet or pass. The dealer places three community cards on the table. Everyone can use these three cards. Players can bet at this point again. The dealer places a fourth card on the table and players can continue to bet. The dealer then places the fifth and final card on the table. A final round of betting takes place. Each player uses their two cards and any three cards on the table to make their best game. The player with the best five cards wins the game. A player's best five cards can be made from any combination of the seven cards available (the five community cards and the two player cards).

Fish

On this simple but interactive game, players start with five cards. The object of the game is to get as many pairs as possible. Whoever ends with the most number of pairs wins. After each player gets their first five cards, he or she discards any pairs on their hand. After discarding their pairs, each player takes turns fishing one card from the player on their right to try to get another pair. At the end of their turn, a player must take additional cards from the center of the deck to maintain five cards on their hand. The game ends when all possible pairs have been formed. The player with the most discarded pairs wins.

I want it

This game plays like fish, however, whoever ends up with the "1" card wins. All cards can be paired up, except the card showing "1". ($\frac{3}{3}$ can be paired with $\frac{2}{2}$, but cannot be paired with the 1 card). Each player gets five cards and discards any pairs. After discarding all pairs, each player fishes from the player on their right until he or she gets a pair. If no pairs are found after fishing from the other players, players continue to fish from the middle deck. Whoever is left holding the "1" card after all pairs have been discarded wins the game.

Concentration

This is the classic memory game. Ten pairs of cards showing the same value are selected. These 20 cards are laid out, face down, in a 4 by 5 grid. Each player takes a turn flipping two cards up, if they have the same value, they take this pair and get one point. They keep flipping two cards as long as they flip a pair. Players try to remember the value and position of the cards that are flipped in order to make new pairs. The player who collects the most pairs wins.

$\frac{1}{2}$	$\frac{2}{4}$	0.5
50%	$\frac{1}{3}$	$\frac{3}{9}$
0.33	$33\frac{1}{3}\%$	$\frac{1}{4}$

$\frac{2}{8}$	$\frac{4}{6}$	0.25
25%	$\frac{2}{3}$	$\frac{3}{4}$
0.66	$66\frac{2}{3}\%$	$\frac{6}{8}$

$\frac{1}{5}$	$\frac{4}{20}$	0.75
20%	$\frac{5}{5}$	$\frac{15}{15}$
0.20	75 %	1.0

$\frac{2}{5}$	$\frac{4}{10}$	0.4
100%	$\frac{9}{15}$	0.6
60 %	40 %	$\frac{3}{15}$

$\frac{4}{5}$	$\frac{8}{10}$	**Wild Card**
80%	$\frac{9}{10}$	$\frac{18}{20}$
0.9	90 %	0.8

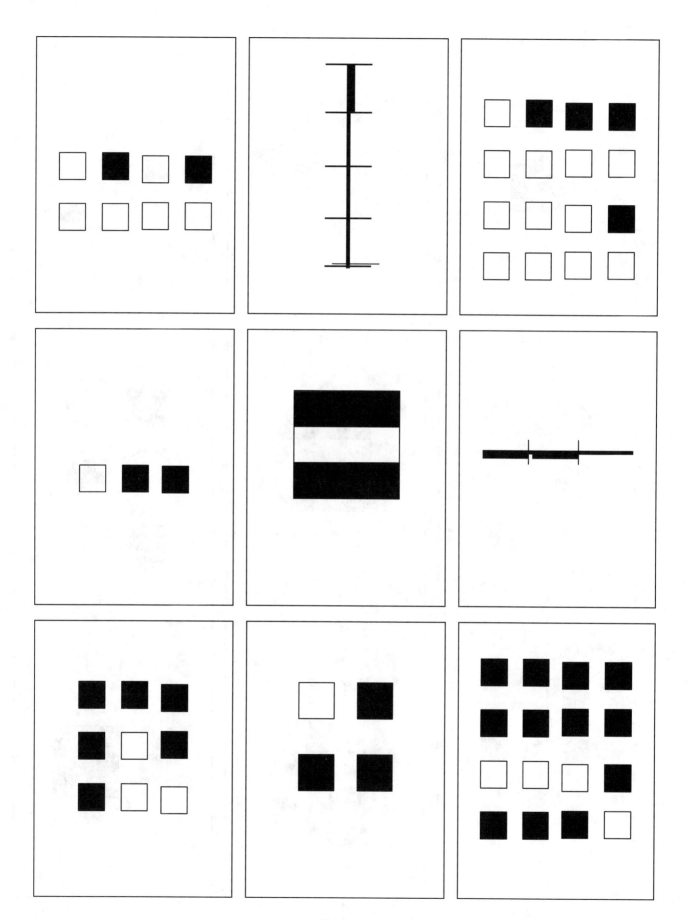

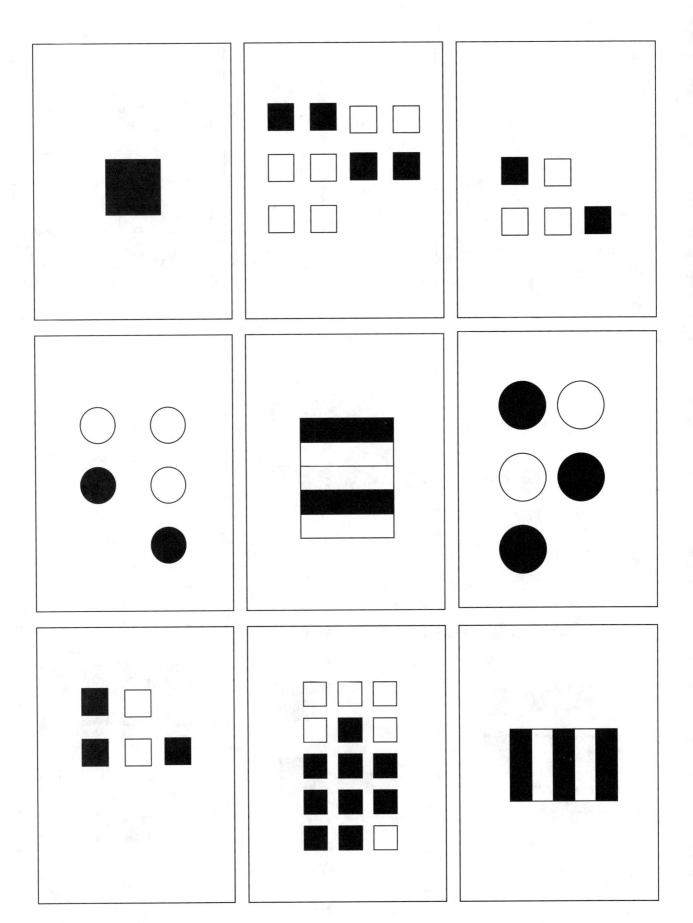

212

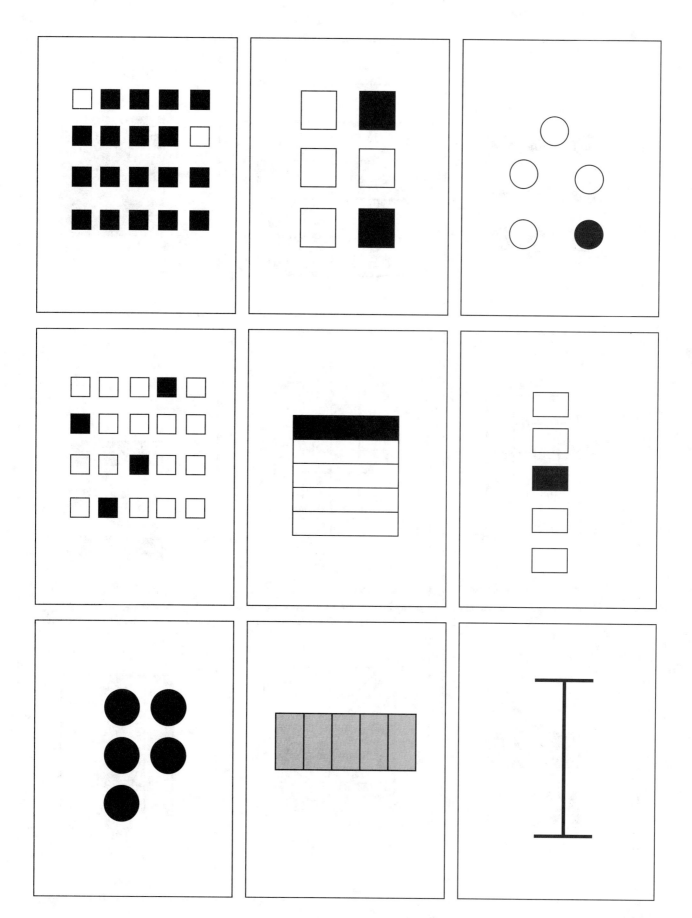

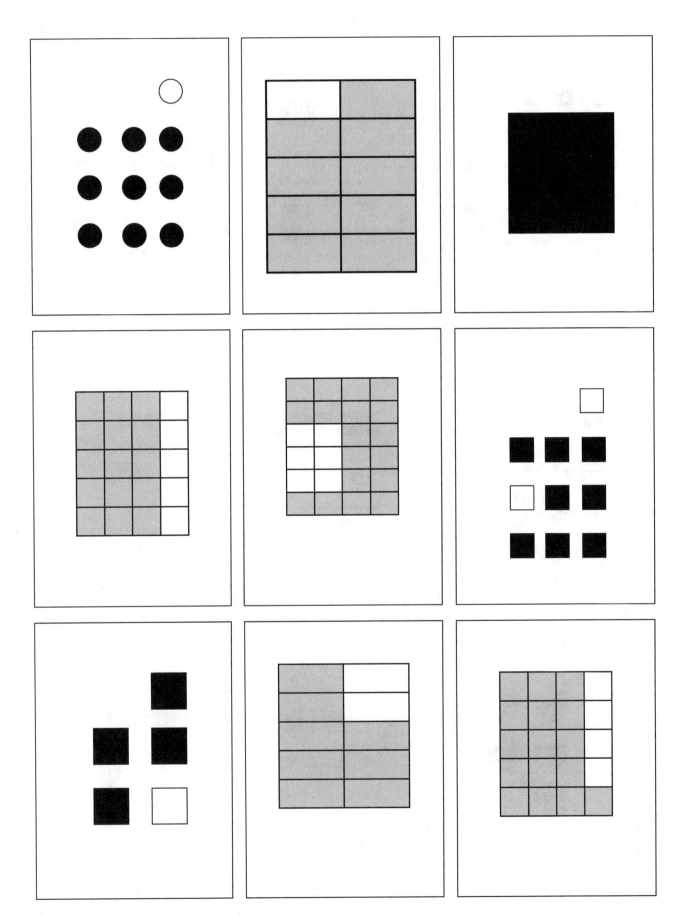

214

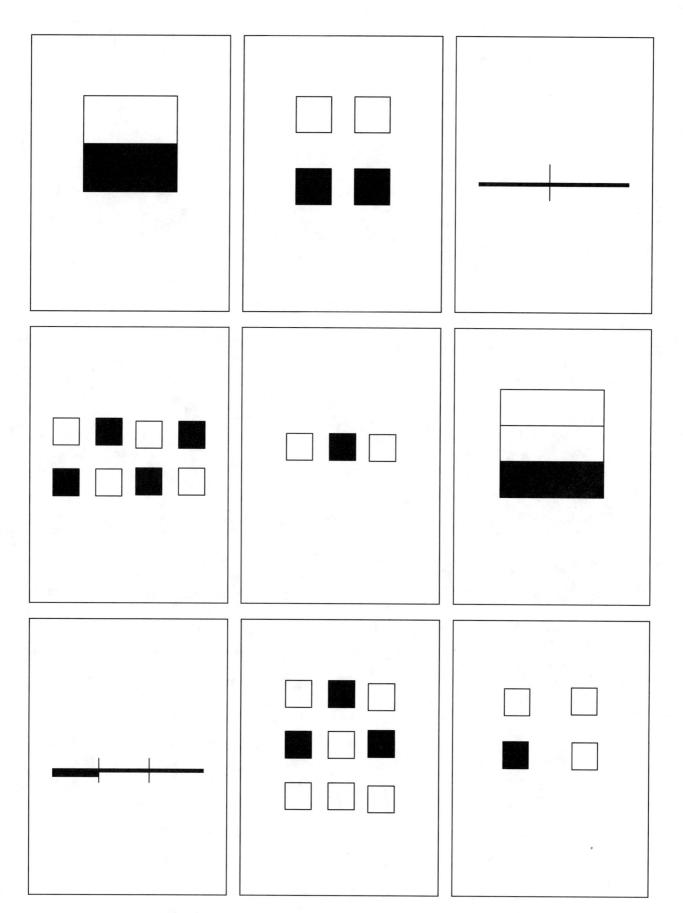

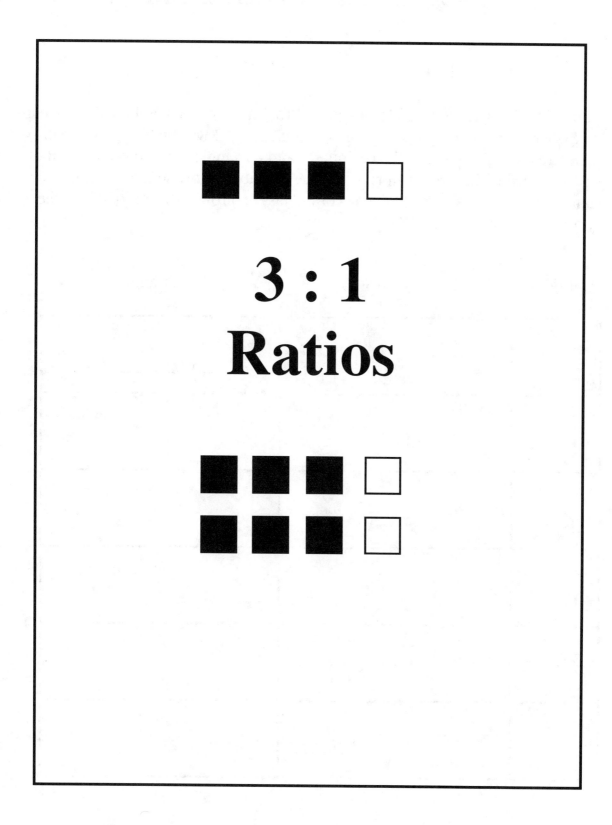

Developing Ratio Understanding (1)

The ratio of squares to triangles above is 1 to 2 (also written as 1:2). This means that when there are 3 figures, 1 will be a square and 2 will be triangles, when there are 6 figures, there will be 2 squares and 4 triangles. Based on the total of figures given in the first column, complete the table below so that the ratio of 1 square to 2 triangles is maintained. For the last two rows, write (not draw) the correct number.

Total Figures	Squares	Triangles
9		△△△△△△
15		
12		
18		
300		
123		

Developing Ratio Understanding (2)

□ □   △ △ △

The ratio of squares to triangles above is 2 to 3 (also written as 2:3). Complete the table below so that the ratio of 2 squares to 3 triangles is maintained.

Squares	Triangles
□ □ □ □	
	△ △ △ △ △ △ △ △ △ △ △ △
□ □ □ □ □ □ □ □ □	
	△ △
□ □ □ □ □ □	
	△ △ △ △ △ △ △ △
□ □ □	

Developing Ratio Understanding (3)

The ratio of horizontal lines to vertical lines above is 3 to 4 (also written as 3:4). Based on the total of lines given in the first column and the ratio shown on the second column (horizontal to vertical ratio), draw the appropriate number of vertical and horizontal lines in the last column of the table below. For the last three rows, write (not draw) the correct number of horizontal and vertical lines.

Total Lines	Ratio	
16	3 : 1	‡ ‡ ‡ ‡ 12 horizontal and 4 vertical lines
12	1 : 5	
15	3 : 2	
16	5 : 3	
21	4 : 3	
64	7 : 1	
100	9 : 1	
128	7 : 9	

Pictorial Mathematics   Ratios

Shading Ratios (1)

Shade-in the grids on the right so that the ratio of shaded to un-shaded spaces matches that of the grid on the left.

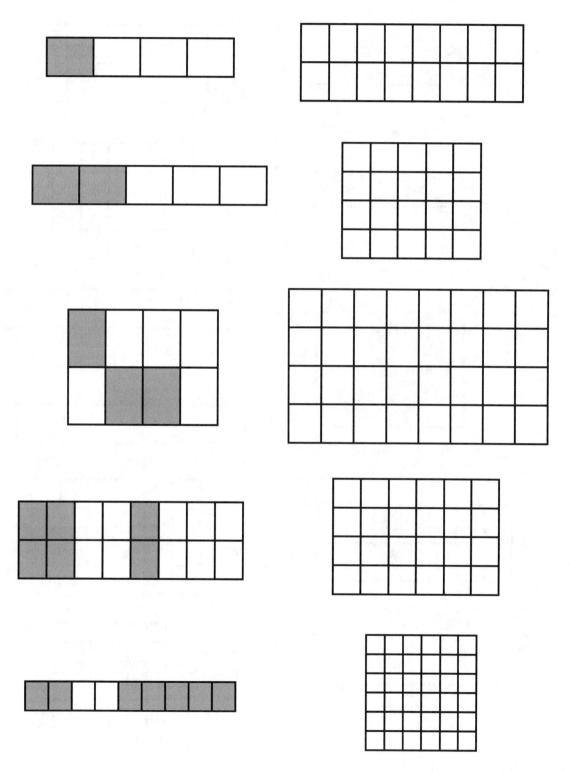

221

Pictorial Mathematics   Ratios

Shading Ratios (2)

Shade-in the grids on the right so that the ratio of shaded to un-shaded spaces matches that of the grid on the left.

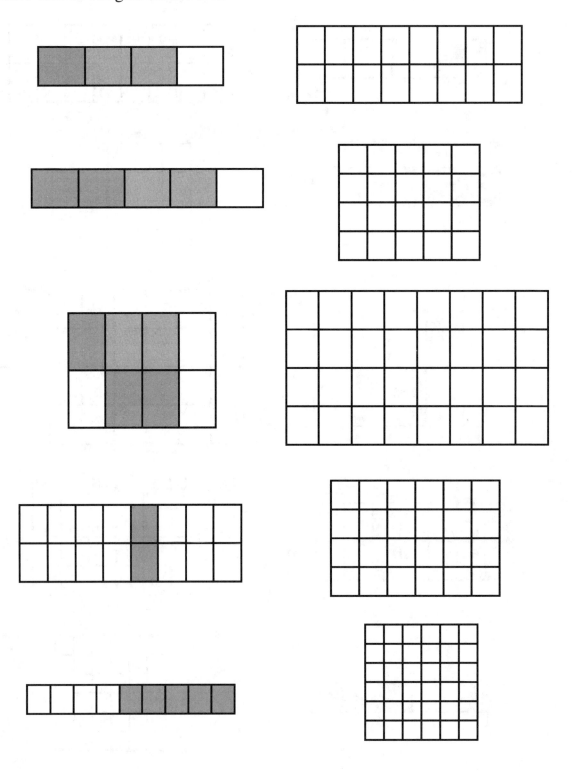

222

Pictorial Mathematics   Ratios

Circle Ratios 1

Shade each of the circles below so that the ratio of shaded to non-shaded parts matches the ratio shown below the circle.

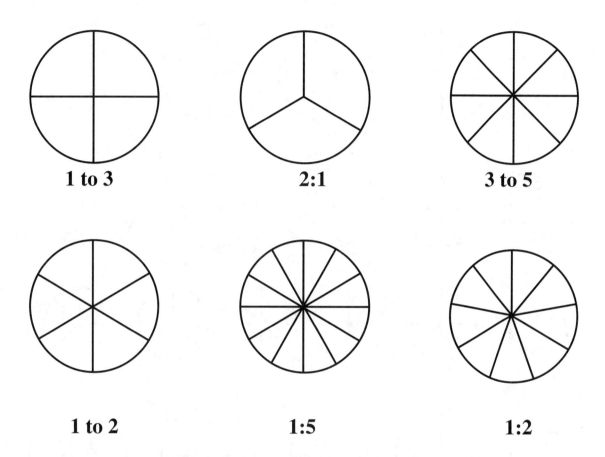

Shade in enough of the six circles below so that the ratio of shaded to non-shaded is 3 to 1.

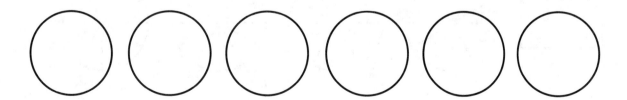

Circle Ratios 2

Shade each of the circles below so that the ratio of shaded to non-shaded parts matches the given ratio.

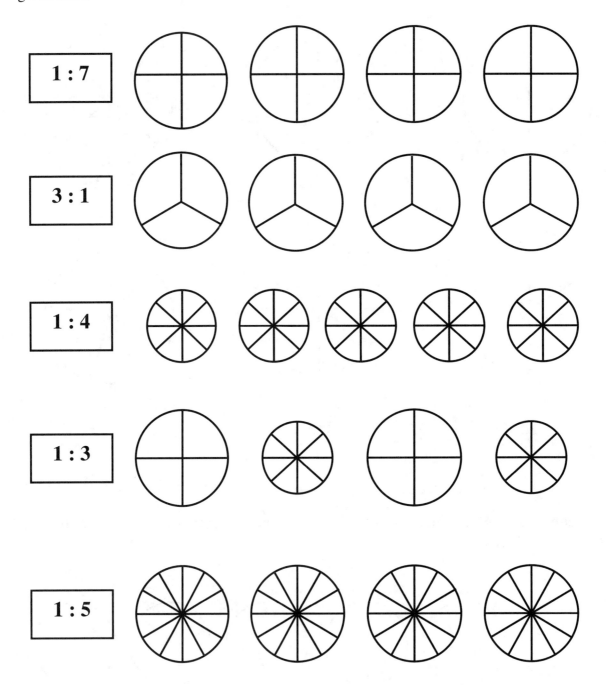

Pictorial Mathematics												Ratios

Developing Ratio Sense (1)

1. What is the ratio of shaded to non-shaded space?

2. Shade the rectangle above so that the ratio of shaded to non-shaded is 3 to 1.

3. What is the ratio of shaded to non-shaded space on the figure below?

4. Shade the rectangle above so that the ratio of shaded to non-shaded is 7 to 8.

5. What is the ratio of shaded to non-shaded space on the figure below?

6. Shade the rectangle above so that the ratio of shaded to non-shaded is 5 to 3.

7. What is the ratio of gray to black rectangles on the figure below?

8. What is the ratio of white to black rectangles?_____. Shade enough space so the ratio of shaded to non-shaded is 2 to 1.

Developing Ratio Sense 2

1. What is the ratio of shaded to non-shaded space?

2. Shade the rectangle above so that the ratio of shaded to non-shaded is 3 to 1.

3. What is the ratio of shaded to non-shaded space on the figure below?

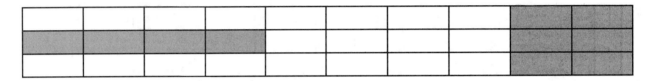

4. Shade the rectangle above so that the ratio of shaded to non-shaded is 7 to 8.

5. What is the ratio of shaded to non-shaded space on the figure below?

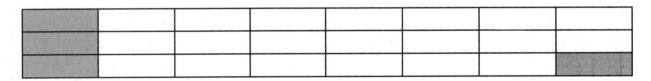

6. Shade the rectangle above so that the ratio of shaded to non-shaded is 5 to 3.

7. What is the ratio of shaded to non-shaded space on the figure below?

8. What fraction of the non-shaded rectangles needs to be shaded so the ratio of shaded to non-shaded rectangles becomes 4 to 5?

Pictorial Mathematics — Ratios

Grid Ratios (1)

Shade each of the rectangles below so that the ratio of shaded to non-shaded space matches the ratio shown below each figure.

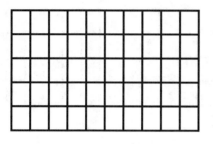

2 : 3   5 : 4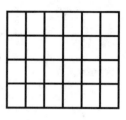

1 : 7

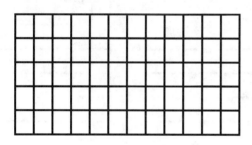

3 : 7   1 : 8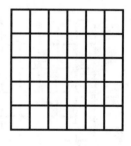

5 : 1

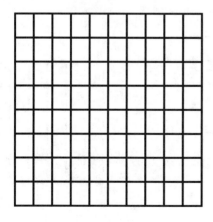

7 : 1   5 : 3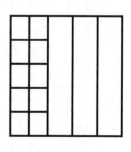

3 : 5

Grid Ratios (2)

Shade each of the rectangles below so that the ratio of shaded to non-shaded space matches the ratio shown below each figure.

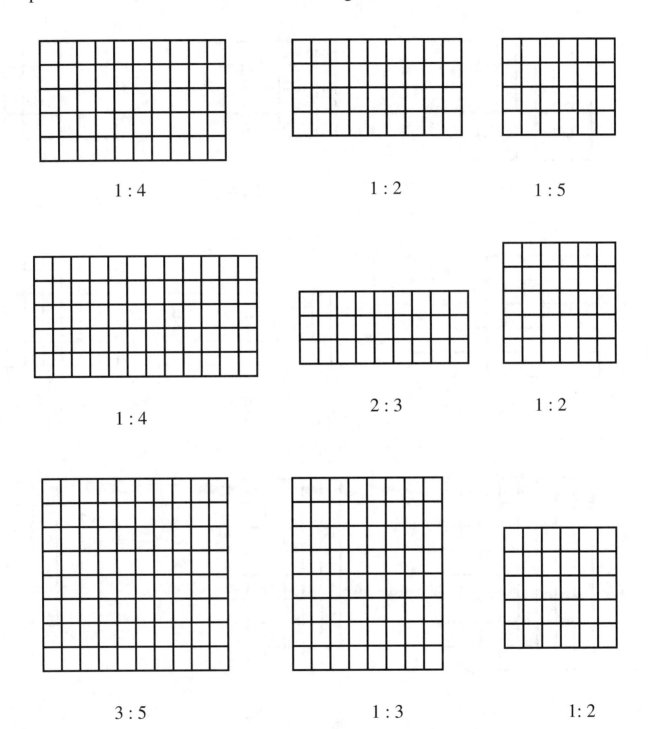

Grid Ratios (3)

Shade each of the rectangles below so that the ratio of shaded to non-shaded space matches the ratio shown below each figure.

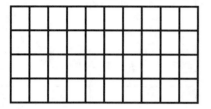
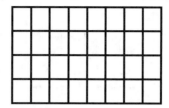
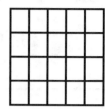

1 : 3   3 : 5   1 : 4

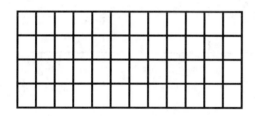
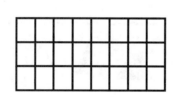
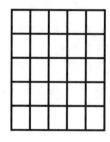

5 : 3   1 : 5   2 : 3

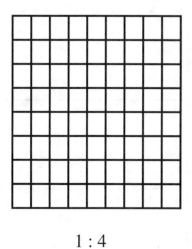
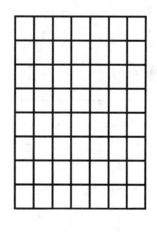
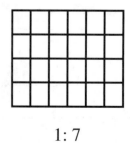

1 : 4   3 : 4   1 : 7

Pictorial Mathematics   Ratios

Ratios vs. Fractions (1)

	Fraction shaded	Ratio of shaded to unshaded
■■■□□□□□□□	$\frac{3}{10}$	**3 : 7**
■□□□□□□□		
△△▲▲▲▲▲▲▲		
△△▲▲▲△△		
○○○○○○○○○● ●●●●●		
●●●○○○○○● ○○●●●○○○●		
△▲ ●○○○○		
△▲▲△△▲▲ ▲▲△▲△		

230

Pictorial Mathematics — Ratios

Ratios vs. Fractions (2)

	Fraction shaded	Ratio of shaded to unshaded
■■■■□□□□		
□■■■■■□□		
▲△▲△▲△▲▲		
△▲▲▲△▲▲		
●●○○○○○○○ ●●●●●●		
○○●○○○○○ ○○●●●●●●		
▲▲ ●●●○○		
▲▲△△△△△▲ ▲△▲△▲△		

231

Ratios vs. Fractions (3)

	Fraction shaded	Ratio of shaded to unshaded
■■■■□□□□□□		
■■■■■□□		
▲△▲△▲△▲		
△▲▲▲△▲		
●○○○○○○ ●●●●		
○○●○○○○ ○○●●●●●●●		
▲▲ ●●○○○		
▲▲ △△▲▲▲▲ ▲△▲△▲		

Pictorial Mathematics   Ratios

Pictorial Transformations (1)

Shade-in the picture on the left in the same proportion as the picture on the right.

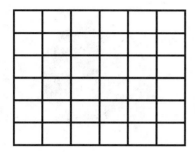 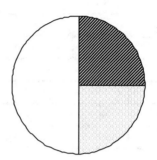

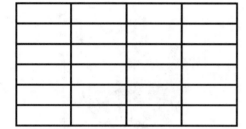 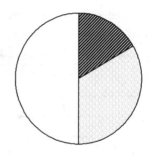

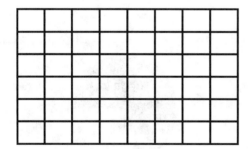 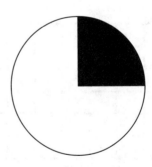

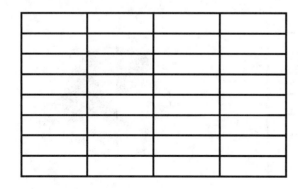 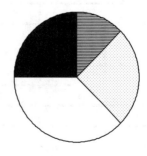

233

Pictorial Mathematics · Ratios

Pictorial Transformations (2)

Shade-in the picture on the left in the same proportion as the picture on the right.

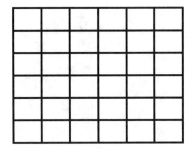 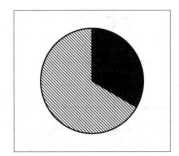

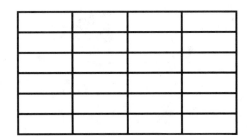 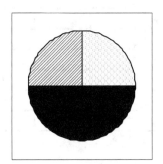

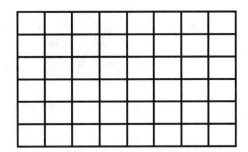 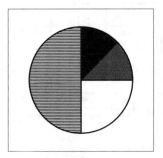

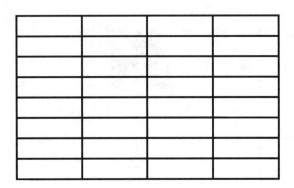 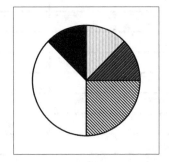

ALGEBRA

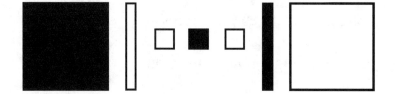

Algebra

What Will Students Be Conceptualizing/Practicing?

- Representing expressions pictorially

- Translating pictorial expressions to standard algebraic symbols

- Transforming simple expressions

- Translating expressions across written, algebraic, and pictorial representations

- Adding expressions pictorially

- Subtracting expressions pictorially

- Representing length as an algebraic expression

- Representing area as an algebraic expression

- The distributive property

- Factoring quadratic monomials in the context of area

- Multiplying expressions

- Multiple representations of zero

- Factoring quadratic trinomials (using only addition)

- Factoring quadratic trinomials (using addition and subtraction)

- Translating quadratics from the standard algebraic representation to a pictorial representation

- Transforming pictorial quadratics by factoring

- Seeing factoring as re-arranging blocks into a rectangular form

- Using variables as referents for the length of line segments

- Identifying points on the number line

- Matching linear equations with their graphs

- Using the graph of y = x as a referent for other linear graphs

- Identifying multiple representations of equivalent expressions

Expression Representations (1)

▭ = x ▬ = $-x$ □ = 1   ■ = −1

Words	Algebraic	Pictorial	Equivalent
One more than a number	$x+1$	▭ □	▭ □ □ ■
One minus a number	$1-x$		
	$2x$	▭ ▭	
	$2x-1$		
	$2x+1$		
	$3x$		
	$3x+1$		
	$1-3x$		
	$4+3x$		

Pictorial Mathematics — Algebra

Expression Representations (2)

$x = \square \quad -x = \blacksquare \quad 1 = \square \quad -1 = \blacksquare$

Words	Algebraic	Pictorial	Equivalent
The sum of a number and 5	x + 5	□ ∷ ∷	☰ ∷ ∷
		∷ ■	
			☰ ∷ ■■
The difference between a number and four			
	3x + 2		
		☰ ∷	
			▬▬▬ ∷
Six less than twice a number			
	x − 8		

239

Pictorial Mathematics   Algebra

Expression Representations (3)

$x^2 = \square$ $-x^2 = \blacksquare$ $\square = x$ $\blacksquare = -x$ $\square = 1$ $\blacksquare = -1$

Words	Algebraic	Pictorial	Equivalent
One more than the square of a number		$\square \; \square$	
The difference between one and the square of a number	$1 - x^2$		
		$\blacksquare \; \square$	
			$\square \; \square \; \blacksquare \; \square$
	$x + x^2$		
			$\square \square \square \; \square \square \blacksquare \square \blacksquare$
The sum of two consecutive numbers			
	$3(x+1)$		
		$\square \; \square \; \square$	

240

Pictorial Mathematics — Algebra

Expression Representations (M)

x^2 = ☐ $-x^2$ = ■   ▭ = x ▬ = $-x$ ▫ = 1   ▪ = -1

Words	Algebraic	Pictorial	Equivalent

Representing Unknowns

A card with a rectangular arrangement of dots is partially covered with an index card. Using the information that is visible, write the expression that describes the total number of dots.

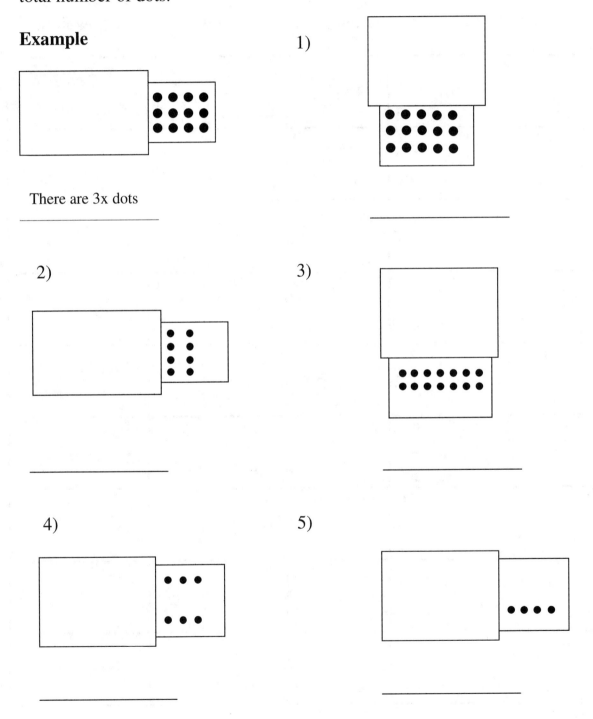

Example

There are 3x dots

Pictorial Algebraic Representations (1)

☐ = x^2 ▭ = x ▫ = 1

Symbolic Representation	Pictorial Representation
$x^2 + 2x + 1$	☐ ▭▭ ▫
$2x^2 + x$	
	☐☐☐ ▭ ▫▫▫▫▫
$4x + 3$	
	☐☐ ▭▭▭ ▫▫▫▫▫▫▫▫▫
$3x^2 + 4$	

Pictorial Algebraic Representations (2)

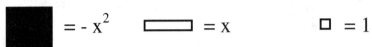

Symbolic Representation	Pictorial Representation
$x^2 - 2x + 1$	
$-2x^2 + x$	
$-4x + 3$	
$-3x^2 - x - 1$	

Pictorial Algebraic Representations (M1)

☐ = x^2 ▭ = x ▫ = 1

Symbolic Representation	Pictorial Representation

Pictorial Algebraic Representations (M2)

☐ = x^2 ■ = $-x^2$ ▭ = x ▫ = 1 ▬ = $-x$ ▪ = -1

Symbolic Representation	Pictorial Representation

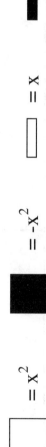

Adding Expressions (2)

If expression 1 + expression 2 = expression 3, draw the missing expressions.

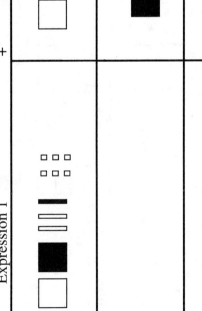

Adding Expressions (M)

☐ = x^2 ■ = $-x^2$ ▭ = x ▬ = $-x$ □ = 1   ■ = -1

If expression 1 + expression 2 = expression 3, draw the missing expressions.

Expression 1	+	Expression 2	=	Expression 3

Subtracting Expressions (1)

□ = x^2 ■ = $-x^2$ | = x] = $-x$ □ = 1 ■ = -1

If expression 1 - expression 2 = expression 3, draw the missing expressions.

Expression 1	Expression 2	Expression 3

Subtracting Expressions (2)

□ = x^2 ■ = $-x^2$ | = x ‖ = $-x$ ▫ = 1 ▪ = -1

If expression 1 − expression 2 = expression 3, draw the missing expressions.

| Expression 1 | − | Expression 2 | = | Expression 3 |

Subtracting Expressions (M)

☐ = x^2 ■ = $-x^2$ [] = x] = $-x$ □ = 1 ▪ = -1

If expression 1 - expression 2 = expression 3, draw the missing expressions.

Expression 1	−	Expression 2	=	Expression 3

Algebraic Number Line (1)

Show where a + 1 and a - 1 could be on the number line below

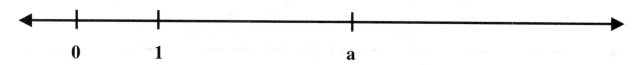

Write-in the approximate value of the point not labeled in terms of the values of a and b.

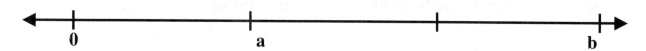

Use the number line below to estimate each of the expressions given in the chart below. Try to make your estimate within 1 unit of the actual value of the expression. Explain your reasoning.

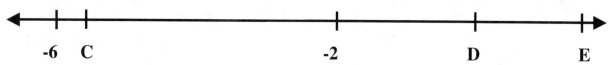

Expression	Value	Reasoning
$-C$		
$C + E$		
DC		
$E - C$		

Algebraic Number Line (2)

For each fraction given, indicate which letter on the number line is the closest to the value of the fraction. Explain your reasoning.

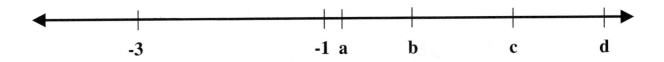

Fraction	Letter Closest to fraction	Reasoning
$\dfrac{18}{19}$		
$-\dfrac{3}{100}$		
$-\dfrac{13}{16}$		

Locate each of the following values on the number line below: 16a, 4a, -2a, a, 9a

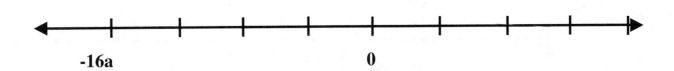

Explain your reasoning for the placement of each point.

Pictorial Mathematics — Algebra

Algebraic Number Line (3)

For each fraction given, indicate which letter on the number line is the closest to the value of the fraction. Explain your reasoning.

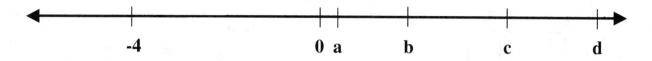

Fraction	Letter Closest to fraction	Reasoning
$\dfrac{19}{9}$		
$\dfrac{33}{8}$		
$\dfrac{47}{8}$		

Locate each of the following values on the number line below: 7a, 3a, -2a, a, -9a

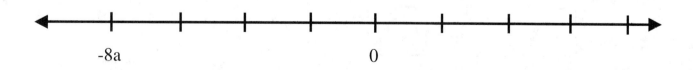

Explain your reasoning for the placement of each point.

Pictorial Mathematics   Algebra

Algebraic Area (1)

Find the area of each of the rectangles (use square units, units^2)

1)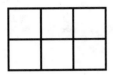

Area: _____

2) 8 / 3

Area: _____

3)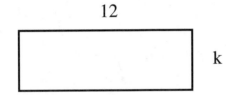

Area: _____

4)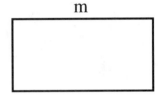

Area: _____

5)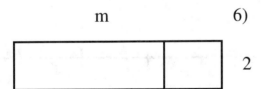

Area: _____

6)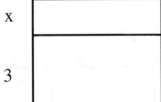

Area: _____

Pictorial Mathematics   Algebra

Algebraic Area (2)

Find the area of each of the rectangles (use square units)

1)   a   3

                     6

Area: _____

2)   b   4

     b

Area: _____

3) $a-2$ 5

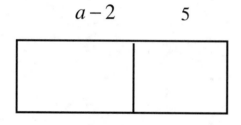

                    a

Area: _____

4)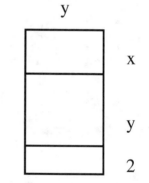

Area: _____

5) $15-a$ 3

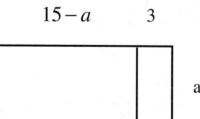

                   a

Area: _____

6)   7   b

    3

    a

Area: _____

| Pictorial Mathematics | Algebra |

Algebraic Area (3)

The rectangle below has a width of 2x and a length of x. It's area is therefore (2x)(x) or $2x^2$. Draw a rectangle for each of the areas given. Label their width and length with their dimensions. Do not use 1x for any of their dimensions. It's up to you to choose the length of x.

2x

x

$3x^2$ $4x^2$

$6x^2$ $5\frac{1}{4}x^2$

Pictorial Mathematics   Algebra

Algebraic Area (4)

The rectangle below has a width of 2x and a length of x. It's area is (2x)(x) or $2x^2$. Draw a rectangle for each of the areas given. Label their width and length with their dimensions. Do not use 1x for any of their dimensions. It's up to you to decide the length of x.

2x

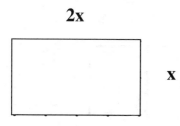

x

$7x^2$ $2\frac{1}{2}x^2$

$1\frac{1}{2}x^2$ $\frac{1}{2}x^2$

Pictorial Mathematics   Algebra

Algebraic Area (M1)

Find the area of each of the rectangles (use square units)

1)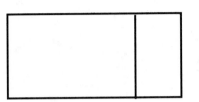

Area: _____

2)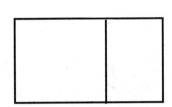

Area: _____

3)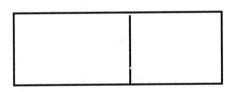

Area: _____

4)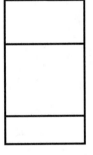

Area: _____

5) 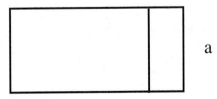   a

Area: _____

6)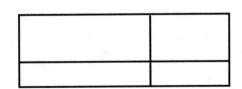

Area: _____

Pictorial Mathematics   Algebra

Algebraic Area (M2)

Draw one rectangle for each of the six areas given. Label their width and length with their dimensions.

Multiplying Expressions (1)

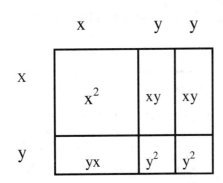

Multiplication shown: $(x + y)(x + 2y)$

Multiplying using distributive property

$(x + y)(x + 2y) = x(x + 2y) + y(x + 2y)$
$= x^2 + 2xy + yx + 2y^2$
$= x^2 + 3xy + 2y^2$

Product shown in the picture: $x^2 + 3xy + 2y^2$

Using the example above, complete the missing parts for each multiplication.

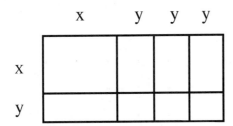

Multiplication shown:

Multiplying using distributive property

Product shown in the picture:

Multiplication shown:

Multiplying using distributive property

Product shown in the picture:

Multiplication shown:

Multiplying using distributive property

Product shown in the picture:

Multiplying Expressions (2)

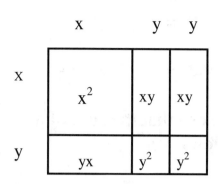

Multiplication shown: $(x + y)(x + 2y)$

Multiplying using distributive property

$$(x + y)(x + 2y) = x(x + 2y) + y(x + 2y)$$
$$= x^2 + 2xy + yx + 2y^2$$
$$= x^2 + 3xy + 2y^2$$

Product shown in the picture: $x^2 + 3xy + 2y^2$

Using the example above, complete the missing parts for each multiplication.

Multiplication shown:

Multiplying using distributive property

Product shown in the picture:

Multiplication shown:

Multiplying using distributive property

Product shown in the picture:

Multiplication shown:

Multiplying using distributive property

Product shown in the picture:

Pictorial Mathematics   Algebra

Shading-in Expressions (1)

☐ = x^2 ■ = $-x^2$ ▭ = x ▫ = 1

▬ = $-x$ ▪ = -1

Shade in the appropriate number of boxes so that the sum of all shaded and un-shaded boxes equals the expression given.

$x^2 + 2x + 4$

$-x^2 + 4x - 2$

$3x^2 - 2x$

$-3x^2$

$x^2 + 6 - 4x$

$4x - 6 - 3x^2$

264

Pictorial Mathematics Algebra

Shading-in Expressions (2)

☐ = x^2 ■ = $-x^2$ ▭ = x ☐ = 1

▬ = $-x$ ■ = -1

Shade in the appropriate number of boxes so that the sum of all shaded and un-shaded boxes equals the expression given.

$-3x^2+2x+2$

x^2+6x-4

$-x^2-2$

x^2

$4+5x^2$

$3x^2-2x$

265

Pictorial Mathematics   Algebra

Shading Expressions (3)

☐ = x^2 ■ = $-x^2$ ▭ = x ▫ = 1

▬ = $-x$ ▪ = -1

Shade in the appropriate number of boxes so that the sum of all shaded and un-shaded boxes equals the expression given.

$-4x^2 + 4x + 2$ ☐ ☐ ☐ ☐ ☐ ☐ ☐ ☐   ▭▭▭▭▭▭▭▭   ▫▫ ▫▫ ▫▫ ▫▫

$x^2 + 6x - 2$ ☐ ☐ ☐ ☐ ☐   ▭▭▭▭▭▭   ▫▫ ▫▫ ▫▫

$-x^2 + 2$ ☐ ☐ ☐ ☐ ☐   ▭▭▭▭▭▭   ▫▫ ▫▫ ▫▫

$3x^2$ ☐ ☐ ☐ ☐ ☐   ▭▭▭▭▭▭   ▫▫ ▫▫ ▫▫

$2 + 5x^2$ ☐ ☐ ☐ ☐ ☐   ▭▭▭▭▭▭   ▫▫ ▫▫ ▫▫

$-3x^2 + 2x$ ☐ ☐ ☐ ☐ ☐   ▭▭▭▭▭▭   ▫▫ ▫▫ ▫▫

Shading-in Expressions (4)

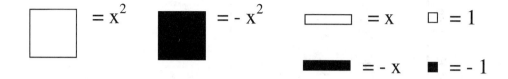

Shade in the appropriate number of boxes so that the sum of all shaded and un-shaded boxes equals the expression given.

$-3x^2 + x + 1$

$5x^2 + 3x - 3$

$-x^2 - 5x + 7$

$-7x^2 + x + 5$

$x - 3 + 5x^2$

Factoring (1)

□ = x^2 ▭ = x ▫ = 1

Symbolic Representation	Pictorial Representation	Factored Pictorial Representation	Factored Symbolic Representation
$x^2 + 2x + 1$	□ ▭▭ ▫	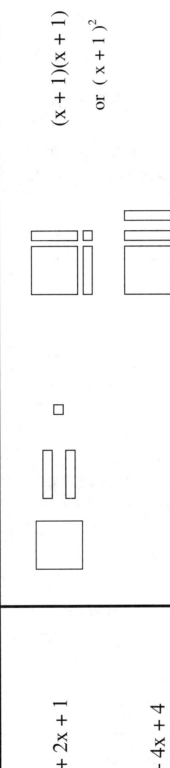	$(x+1)(x+1)$ or $(x+1)^2$
$x^2 + 4x + 4$			
$x^2 + 6x + 9$			
$x^2 + 3x + 2$			

Pictorial Mathematics   Algebra

Factoring (2)

☐ = x^2 ▭ = x ▫ = 1

Symbolic Representation	Pictorial Representation	Factored Pictorial Representation	Factored Symbolic Representation
$x^2 + 5x + 6$			
	▭▭▭▭▭ ▫▫▫	☐▭ ▫	$(x+1)(x+3)$
$x^2 + 6x + 5$			

269

Pictorial Mathematics — Algebra

Factoring (3)

☐ = x^2 ■ = $-x^2$

☐ = x ■ = $-x$ ☐ = 1 ■ = -1

Symbolic Representation	Pictorial Representation	Factored Pictorial Representation	Factored Symbolic Representation
$x^2 - 3x + 2$	☐ ■■■ ☐☐		
$x^2 - x - 6$		☐ ■	$(x-1)(x+3)$

Factoring (4)

□ = x^2 ▭ = x ▫ = 1

Symbolic Representation	Pictorial Representation	Factored Pictorial Representation	Factored Symbolic Representation
$3x^2 + 5x + 2$			
			$(2x+1)(x+3)$
$2x^2 + 6x + 4$			

Pictorial Mathematics — Algebra

Factoring (5)

□ = x^2 ■ = $-x^2$ ▭ = x ▬ = $-x$ ▫ = 1 ▪ = -1

Binomial Expression	Pictorial Binomial	Factored Pictorial	Factored binomial
$x^2 + 2x + 1$	□ ‖ ▫	□‖	$(x+1)(x+1)$
$x^2 + 4x + 4$			
	□ ‖‖ ▫▫		
		□ ‖‖‖	
			$(x+2)(x+3)$
$x^2 - x - 2$			
	□ ▮▮▮ ▫▫▫		

272

Factoring (6)

☐ = x^2 ■ = $-x^2$ ▭ = x ▬ = $-x$ □ = 1 ■ = -1

Symbolic Representation	Pictorial Representation	Factored Pictorial Representation	Factored Symbolic Representation
x^2-2x+1			$(x-1)(x-1)$
x^2-3x+2			
x^2-4x+4			
x^2-5x+6			

Pictorial Mathematics — Algebra

Factoring (M1)

☐ = x^2 ▭ = x □ = 1

Symbolic Representation	Pictorial Representation	Factored Pictorial Representation	Factored Symbolic Representation

Factoring (M2)

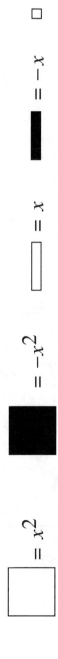

Symbolic Representation	Pictorial Representation	Factored Pictorial Representation	Factored Symbolic Representation

Matching Expressions (1)

Find the two expressions in columns 3 and 4 that match (are equivalent) to each expression in column one. Use column 2 to write the letter of the expressions that are equivalent to the expressions in column 1.

Column 1	Column 2	Column 3	Column 4
$(x-3)(x-1)$		a ▭❘❘❘❘ ∷	i $x^2 + 4x + 4$
▭❘❘❘❘❘ ∷		b $(x+1)(x-1)$	j $(x+4)(x+1)$
▭❘❘❘❘ / ▬▫▫▫▫		c ▭❘❘❘❘ ∷	k $x^2 - 4$
$(x-2)^2$		d $(x-2)(x+2)$	l ▭❘❘
▭❘❘❘❘ ·		e ▭❘❘❘ / ▬▫▫▫	m ▭❘❘ / ▬∷
▭❘ ∶		f $x^2 - 3x - 4$	n $(x+4)(x-1)$
▭❘❘❘ ∷		g $(x-2)(x+2)$	o ▭❘❘❘❘ ∷
$(x+2)^2$		h ▭❘❘❘❘ / ▬▫▫▫▫	p ▭❘❘ / ▬

Pictorial Mathematics   Algebra

Matching Expressions (2)

Find two expressions in columns 3 and 4 that match (are equivalent) to each expression in column one. Use column 2 to write the letter of the expressions that are equivalent to the expressions in column 1.

Column 1	Column 2	Column 3	Column 4
$(x+3)(x+1)$		a [pictorial]	i $x^2 - x - 6$
[pictorial]		b $(x+1)(x+1)$	j $(x-4)(x-1)$
[pictorial]		c [pictorial]	k $x^2 - 9$
$(x-1)^2$		d $(x-2)(x+2)$	l [pictorial]
[pictorial]		e [pictorial]	m [pictorial]
[pictorial]		f $x^2 - 5x + 4$	n $(x+4)(x-1)$
[pictorial]		g $(x-3)(x+3)$	o [pictorial]
$(x-3)(x+2)$		h [pictorial]	p [pictorial]

277

Linear Equation Match (1)

For each linear equation identify which of the four graphs given best matches the equation. Write one or two sentences to explain your selection.

1) $y = -x + 2$

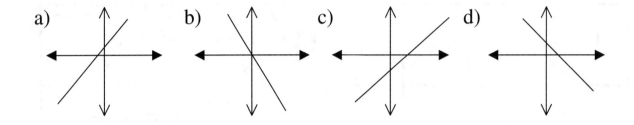

2) $y = \frac{1}{4}x$

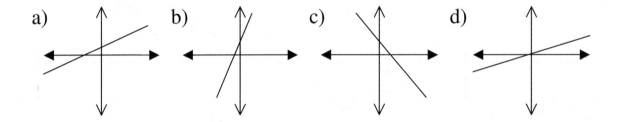

3) $y = x + 1$

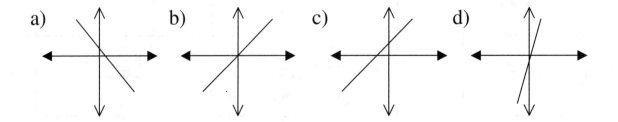

Linear Equation Match (2)

For each linear equation identify which of the four graphs given best matches the equation. Write one or two sentences to explain your selection.

1) $y = 2x$

a) 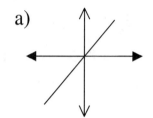 b) 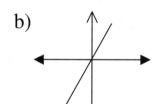 c) 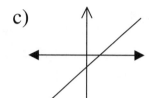 d)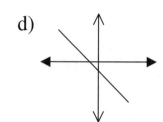

2) $y = -2x - 3$

a) 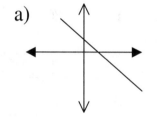 b) 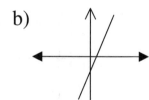 c) 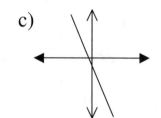 d)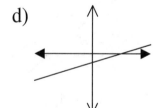

3) $x = -3$

a) 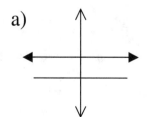 b) 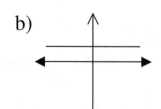 c) 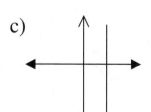 d)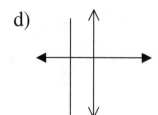

Linear Equation Match (3)

For each linear equation identify which of the four graphs given best matches the equation. Write one or two sentences to explain your selection.

1) $y = -2x$

a) 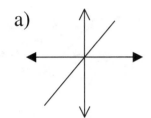 b) 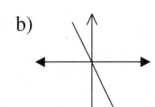 c) 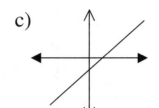 d)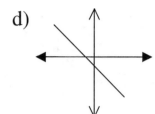

2) $y = -2$

a) 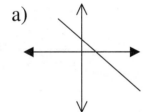 b) 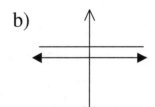 c) 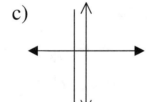 d)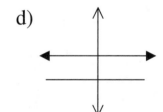

3) $x = 0.5$

a) 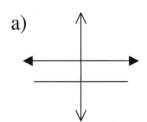 b) 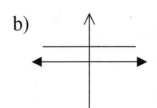 c) 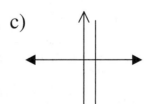 d)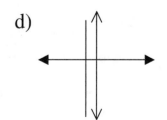

Algebraic Linear Measurement (1)

Each of the lines below is split into sections of various lengths. Find the length of each line. Assume the unit being used is cm.

1) 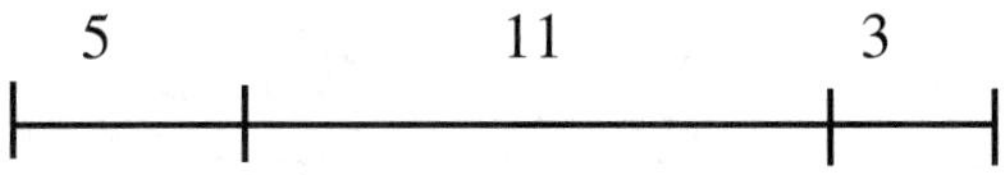
5
11
3

Length: _______________

2) 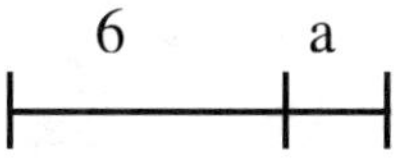
6
a

Length: _______________

3) 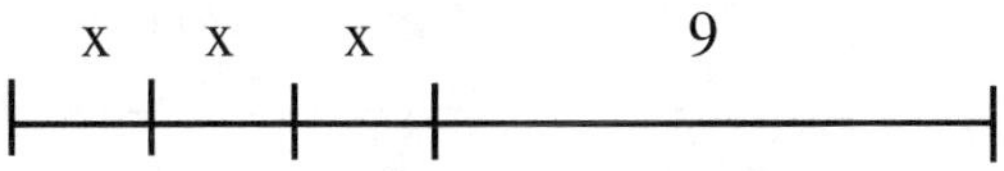
x
x
x
9

Length: _______________

4) 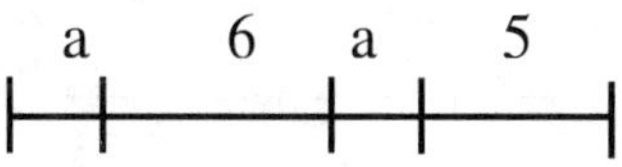
a
6
a
5

Length: _______________

5) 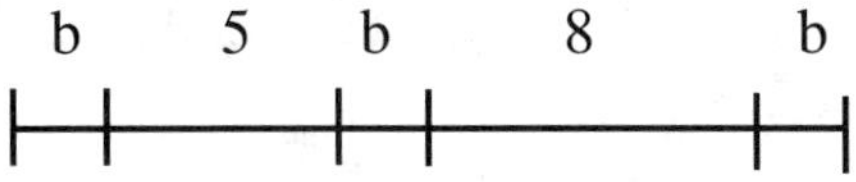
b
5
b
8
b

Length: _______________

6) 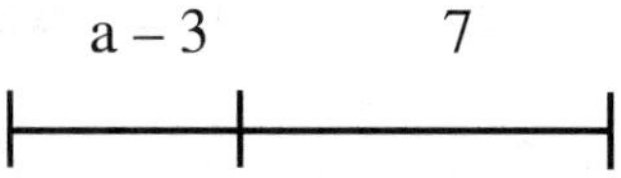
a – 3
7

Length: _______________

7) 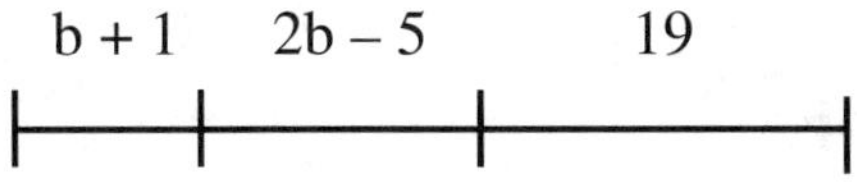
b + 1
2b – 5
19

Length: _______________

8) 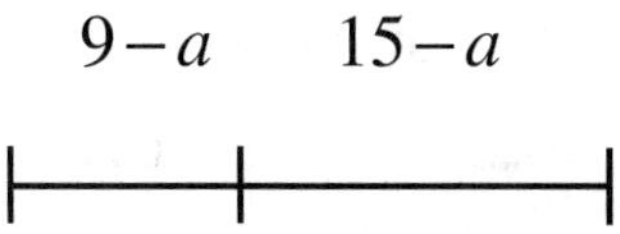
9 – a
15 – a

Length: _______________

Pictorial Mathematics Algebra

Algebraic Linear Measurement (2)

Each of the lines below is split into sections of various lengths. Find the length of of each line. Assume the unit being used is cm.

1) $b+3$ | $3b+3$

Length: _____

2) $a+3$ | $3(a+3)$

Length: _____

3) $8-w$ | $16-w$

Length: _____

4) $8-m$ | $2(8-m)$

Length: _____

5) $r-3$ | $b+2$ | $b+r$

Length: _____

6) $\frac{1}{2}k$ | k | $\frac{3}{4}k$

Length: _____

7) $9-x$ | $y+2$ | $x+y-4$ | 1

Length: _____

8) $\frac{1}{2}k+\frac{3}{4}$ | $\frac{3}{4}k-\frac{1}{8}$

Length: _____

Pictorial Mathematics   Algebra

Algebraic Linear Measurement (M)

Each of the lines below is split into sections of various lengths. Find the length of each line. Assume the unit being used is cm.

1)

   Length: _____

2)

   Length: _____

3)

   Length: _____

4)

   Length: _____

5)

   Length: _____

6)

   Length: _____

7)

   Length: _____

8)

   Length: _____

Algebra Card Games

The following are some examples of the type of games students can play with the algebra cards included in this section to further develop their conceptual understanding. It includes a blank template that teachers and students can use to make their own cards using the sample cards in the set as a guide.

There are 54 algebraic expression cards ($x + 2$, two more than x, etc.). About half of them are pictorial (see samples below). For each expression (i.e. $x - 1$), there are at least three other equivalent expressions. A good way to get students familiar with the equivalent expressions involved is to have them sort them before they start a game.

Algebra Poker

Students play in groups of three to five. The first few times the teacher might play with the whole class by copying a set of two player's cards on transparencies. Each player gets a total of five cards. The object of the game is to make the highest pair, three of a kind, or four of a kind possible. For example:

Player 1

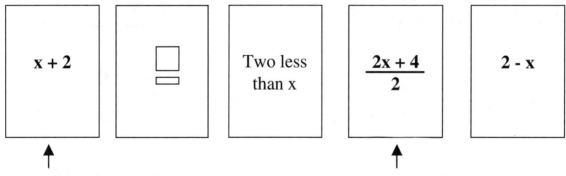

This player has one pair of cards showing the same expression, $x + 2$. He or she will keep this pair, and will ask to have the dealer to change the other three cards to improve his/her game.

Player 2

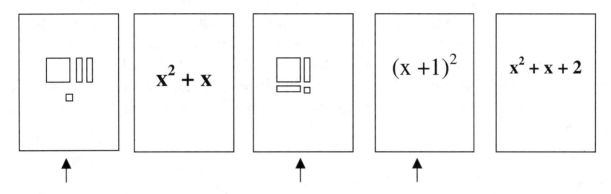

Player 2 has three cards with the same expression, $x^2 + 2x + 1$. This player is said to have three of a kind, which always beats any pair. As it stands right now, player 2 has a better hand than player 1. Here is the order of best possible hands, starting with the worse possible hand, not even one pair, and ending with the best possible hand, four of a kind.

1) **No pair.** When nobody has a pair, the player with the highest card wins. The highest card in the deck is determined by substituting "1" for the value of x into each card's expression.

2) **One pair.** As before, if two players have a pair, the player with highest value pair wins. For example, a pair of halves, beats a pair of thirds

3) **Two pairs**

4) **Three of a kind**

5) **Full house, or three of kind and two of a kind.**

6) **Poker, or four of a kind.**

Poker variations

- Give all players an equal number of chips to bet. Each player puts in 1-3 chips before they get their cards. They can bet after they get their first five cards and they can make a final bet after they exchange cards.

- **Hold them**. Players bet one to three chips before getting any cards. Players get two cards. They bet or pass. The dealer places three community cards on the table. Everyone can use these three cards. Players can bet at this point again. The dealer places another community card on the table and players can continue to bet. The dealer then places the fifth and final card on the table. A final round of betting takes place. Each player uses any combination of their cards and the community cards on the table to make their best five card game.

Fish
On this simple but interactive game, players get five cards. The object of the game is to get as many pairs as possible. Whoever ends with the most number of pairs wins. After each player gets their first five cards, he or she discards any pairs on their hand. After discarding their pairs, each player takes turns fishing one card from the player on their right to try to get another pair. At the end of their turn, a player must take additional cards from the center of the deck to maintain five cards on their hand. The game ends when all pairs have been found. The player with the most discarded pairs wins.

I want it
This game plays like fish, however, whoever ends up with the x^{-1} card wins. All cards can be paired up and discarded when paired, except the card showing x^{-1} (1/x can be paired with the card that reads "the reciprocal of x, but cannot be paired with the x^{-1} card). Each player gets five cards and discards any pairs. After discarding all pairs in hand he or she fishes from other players until he or she gets a pair. If no pairs are found after fishing from the other players, players continue to fish from the middle deck. Everyone wants to have the x^{-1} card because whoever is left holding this card after all pairs have been discarded wins the game.

Concentration
This is the classic memory game. 10 pairs of cards showing the same value are selected. These 20 cards are laid out, face down, in a 4 by 5 grid. Each player takes a turn flipping two cards up, if they have the same expression, they take this pair and get one point. They keep flipping two cards as long as they flip a pair. Players try to remember the value and position of the cards that are flipped in order to make new pairs.

$\dfrac{4x-4}{2}$	$(x+2)^2$	$x^2 + 4x + 4$
$x(x+5) + 6$	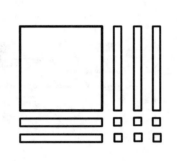	$x^2 + 5x + 6$
$(x+3)(x+2)$	**Subtract a number from 2**	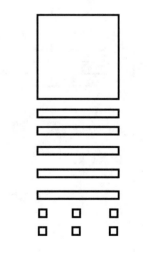

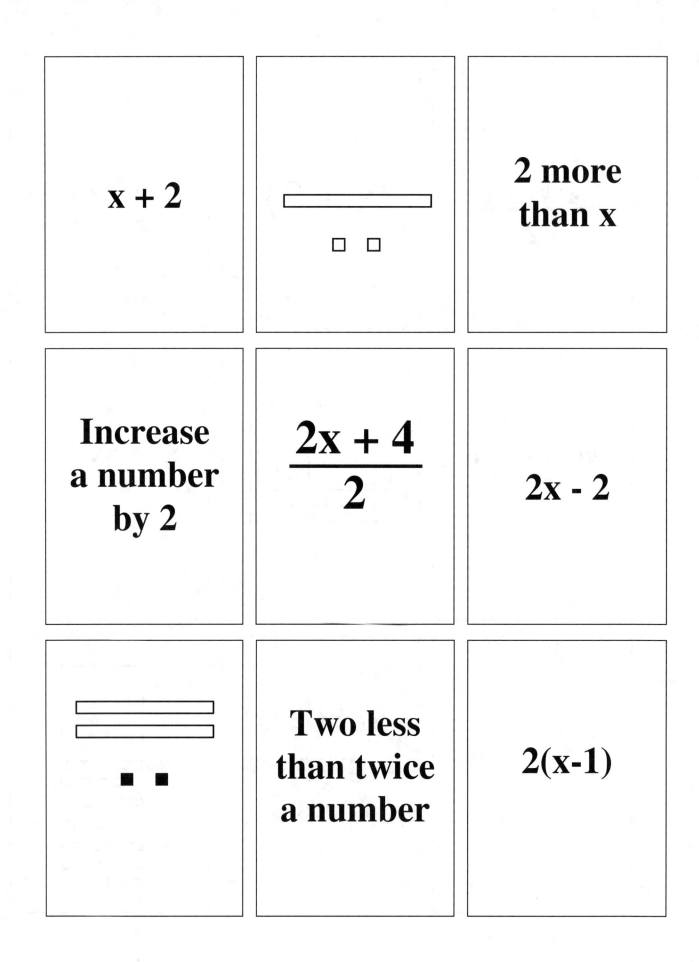

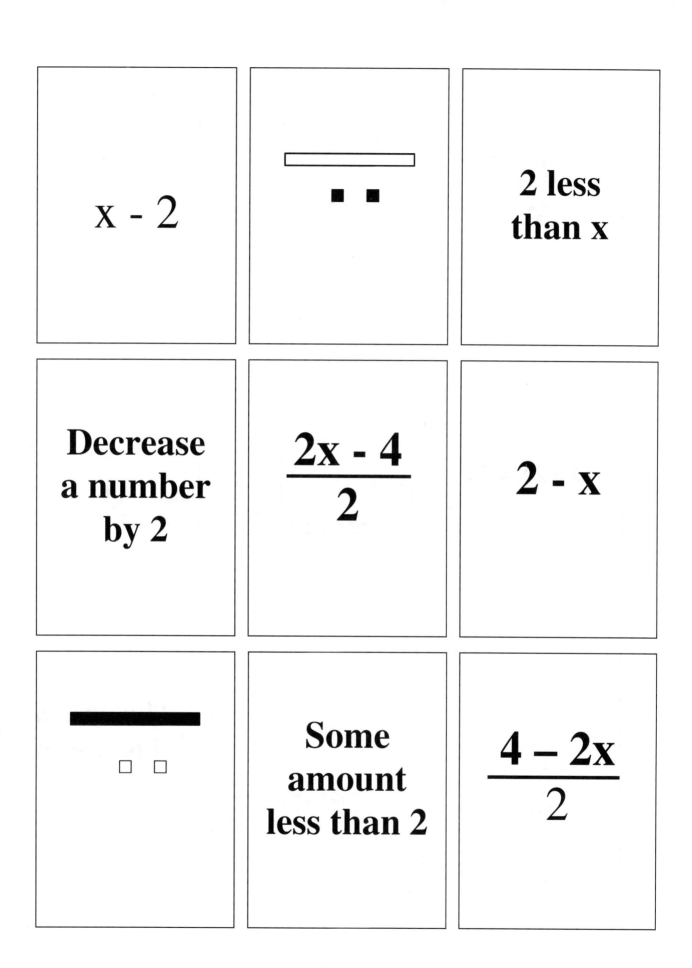

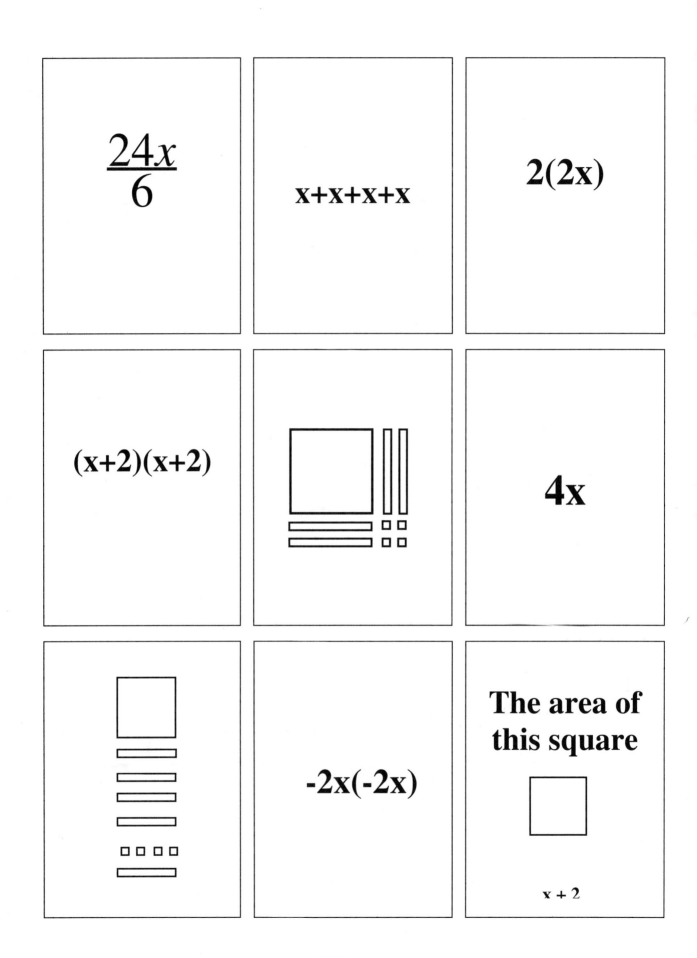

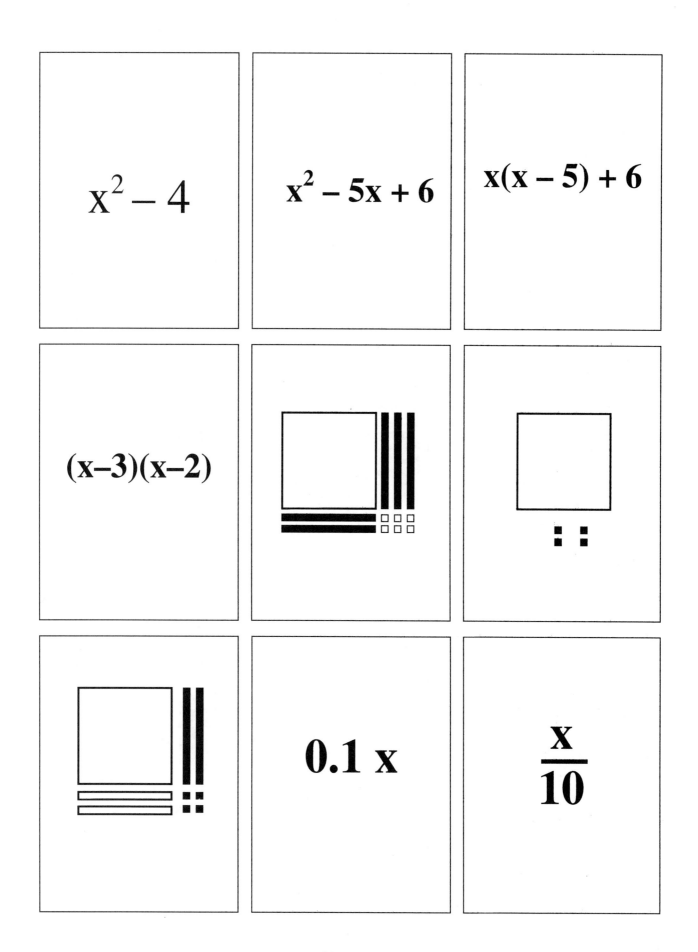

One tenth of a number	$\dfrac{1}{x}$	**The reciprocal of x**
10% of x	x^{-1}	**One divided by x**
x divided by 10	**0.1 x**	$\dfrac{x}{10}$

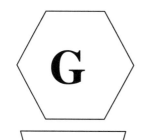

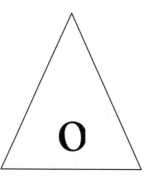

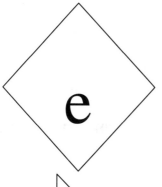
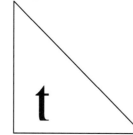
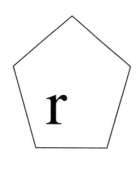
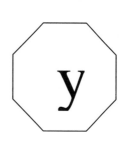

- Perimeter
- Area
- Constructions
- Polygons
- Properties
- Relationships
- Analogies
- Errors
- Tangrams
- Compass
- Pascal's Triangle
- Reflections
- 3-D constructions

Pictorial Mathematics   Geometry

Constructing Figures

Connect the dots to make the given figures

A rhombus

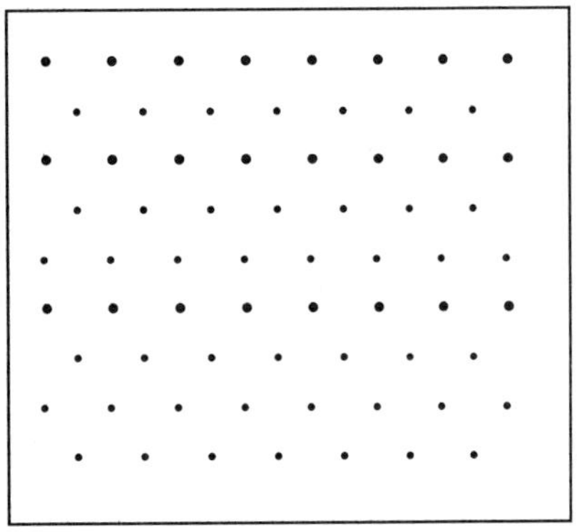

A right triangle

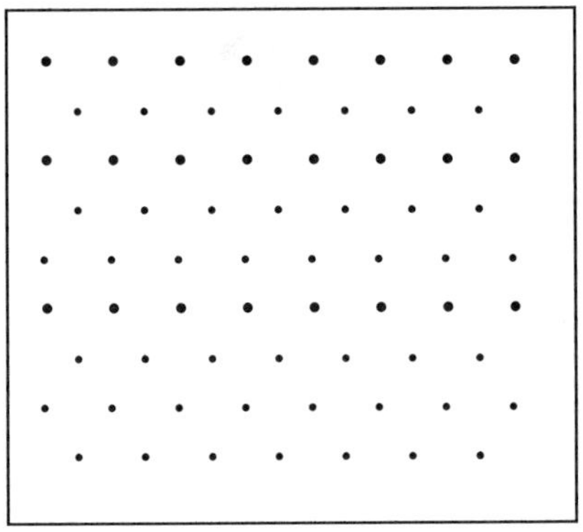

An equilateral triangle

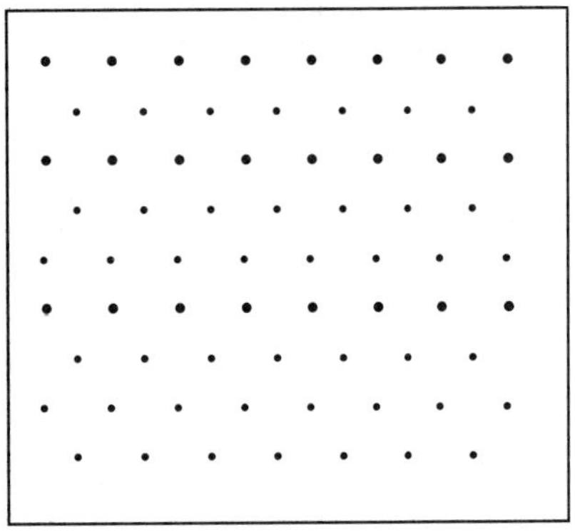

A trapezoid

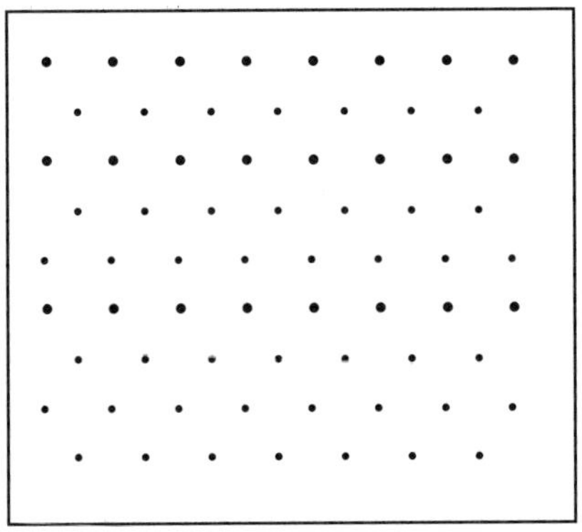

A parallelogram

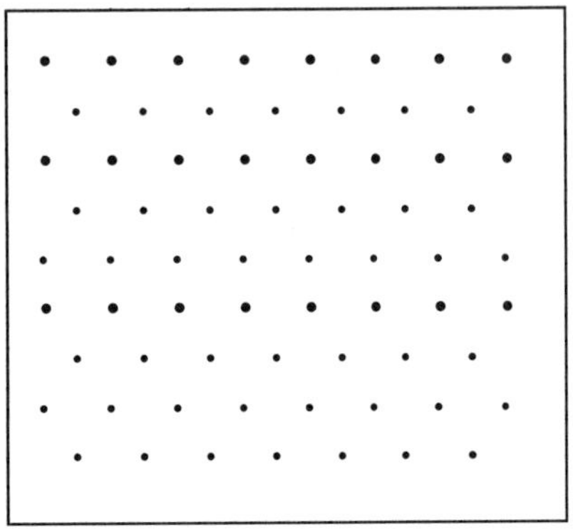

Two congruent triangles

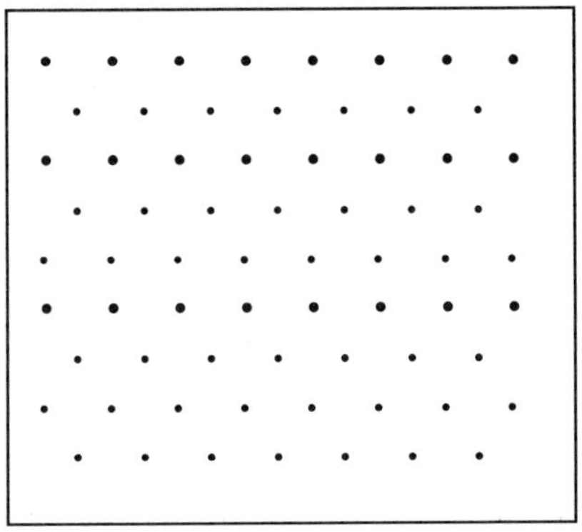

Pictorial Mathematics					Geometry

Area – Perimeter Connections (1)

(assume that the distance from one dot to the next is 1 cm.)

1) Draw all the rectangles with an area of 12 squared cm. Find and write each rectangle's perimeter.

2) Draw all the rectangles with a perimeter of 12 cm. Find and write each rectangle's area.

3) Draw all the rectangles with an area of 18 squared cm. Find and write each rectangle's perimeter.

Area – Perimeter Connections (2)

(assume that the distance from one dot to the next is 1 cm.)

2) Draw all the rectangles with an area of 16 squared cm. Find and write each rectangle's perimeter.

3) Draw all the rectangles with a perimeter of 20cm. Find and write each rectangle's area.

3) Draw all the rectangles with an area of 24 squared cm. Find and write each rectangle's perimeter.

Area – Perimeter Connections (3)

(assume that the distance from one dot to the next is 1 cm.)

1) Draw all the rectangles with a perimeter of 18 cm. Find and write each rectangle's area.

2) Find and draw the rectangle with the largest area whose perimeter is 36 cm. Also find and draw the rectangle with the largest perimeter whose area is 32 cm^2.

Find the Perimeter

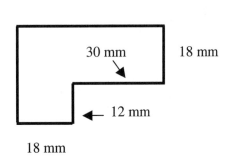

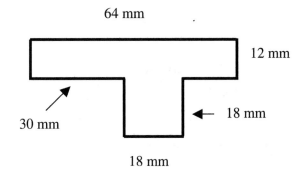

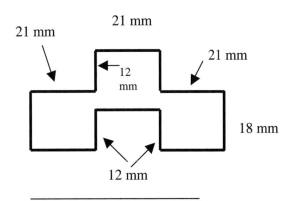

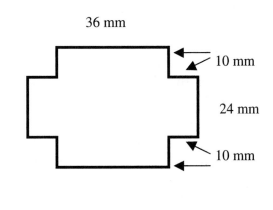

Using a ruler, draw an irregular shape similar to one of the four above so that the perimeter is 230 mm.

Algebraic Perimeter

Write the perimeter for each of the figures

1)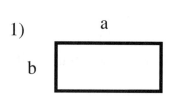

2)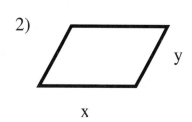

3)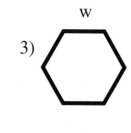

4) a, b, c (trapezoid)

5) 2a + b, b (rectangle)

6)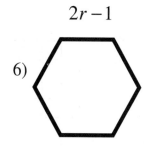

7) 2a + b + 0.5, 2b − a (parallelogram)

8) 8 − 2e, 2e + r, 2r − e (triangle)

301

Draw the Area

Describe your method

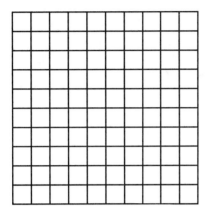

Draw an irregular pentagon with an area between 16 and 19 square units.

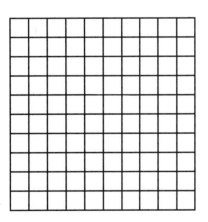

Draw an irregular octagon with an area between 16 and 19 square units.

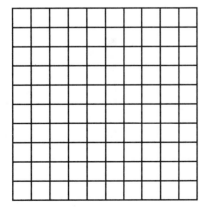

Draw a circle with an area between 16 and 19 square units.

Pictorial Mathematics   Geometry

Find the Area

Show your work
Describe your method

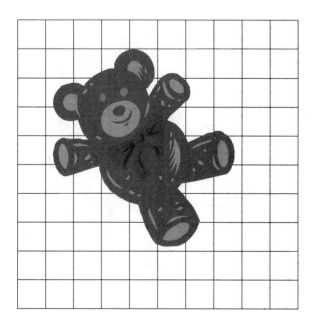

Pictorial Mathematics — Geometry

Find the Area Clip-Art

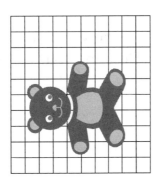  

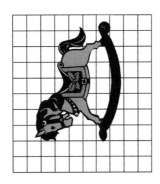 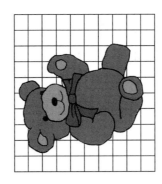 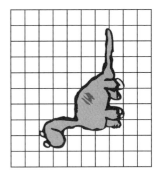

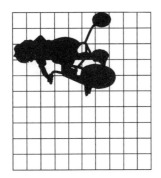  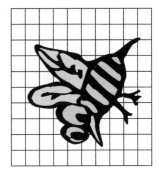

Reflections 1

Draw the reflection of the figure across the given axis.

Example

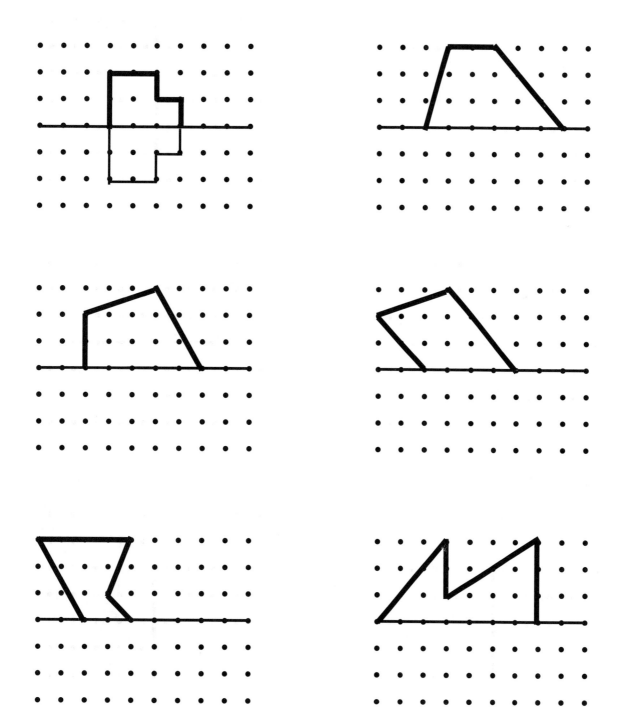

Reflections 2 (M)

Draw the reflection of the figure across the given axis.

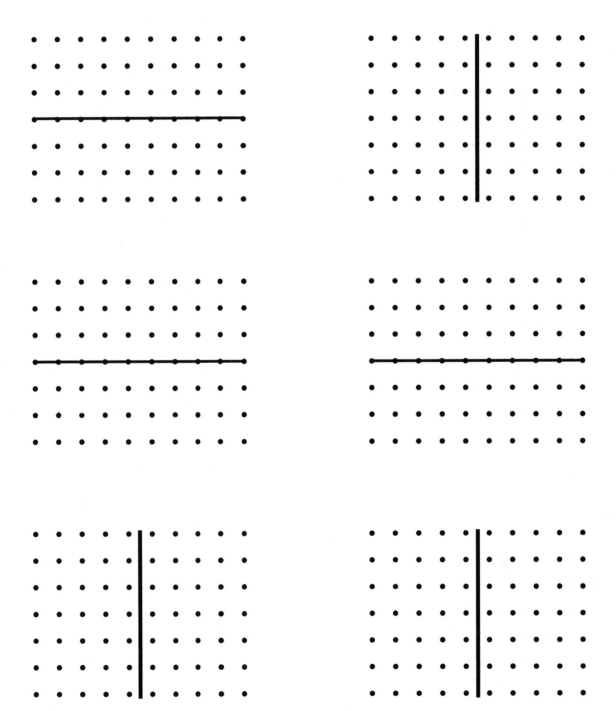

306

Pictorial Mathematics   Geometry

Geometrical Compass (1)

clockwise   Counter-clockwise

Starting at the circle inside the grid, draw the lines following the directions given for each grid. Assume that the distance from one point to another is 1 block, including between diagonal points.

Example

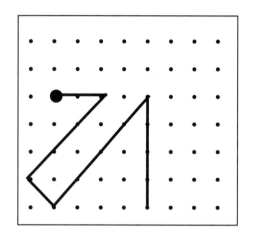

Directions

2 blocks East, 3 blocks SW, 1 block SE, 4 blocks NE, 4 blocks S.

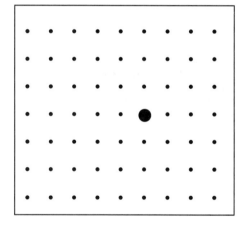

Directions

3 blocks East, 3 blocks SW, 4 blocks W, 5 blocks N, 4 blocks SE.

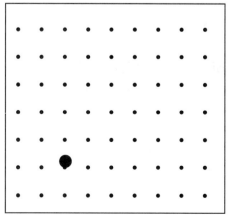

Directions

4 blocks NE, 3 blocks S, 2 blocks W 2 blocks N, 1 block SE, 3 blocks W.

Pictorial Mathematics   Geometry

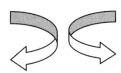 **Geometrical Compass (2)**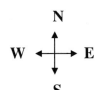

clockwise   Counter-clockwise

Starting at the circle inside the grid, draw the lines following the directions given for each grid. Assume that the distance from one point to another is 1 block, including between diagonal points.

Directions

2 blocks South, 3 blocks SW, 1 block W,
4 blocks N, 2 blocks E, 2 blocks SE.

Directions

4 blocks N, 3 blocks SE, 3 blocks W,
3 blocks NE, 3 blocks, 1 block SE.

Directions

6 blocks NE, 4 blocks S, 3 blocks W
2 blocks SE, 1 block E, 2 blocks N.

308

Pictorial Mathematics   Geometry

Isometric Copies

The ability to visualize 3 dimensional shapes from various perspectives is extremely helpful in a variety of fields such as higher mathematics, architecture, engineering, art, computer drafting, etc. Make an identical copy of each figure below.

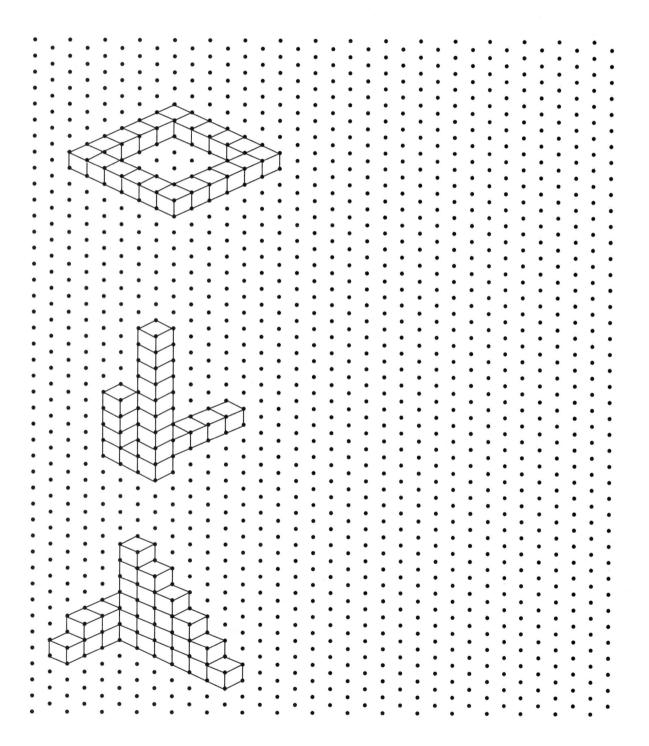

Geometrical Directions

Your best friend was absent and needs you to tell him what homework is due tomorrow. You are talking to your friend over the phone. Describe each of the shapes below so your friend can draw them on her grid paper.

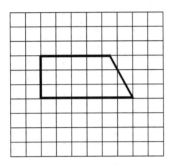

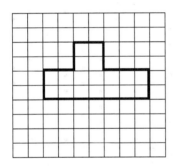

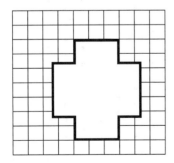

Geometrical Analogies

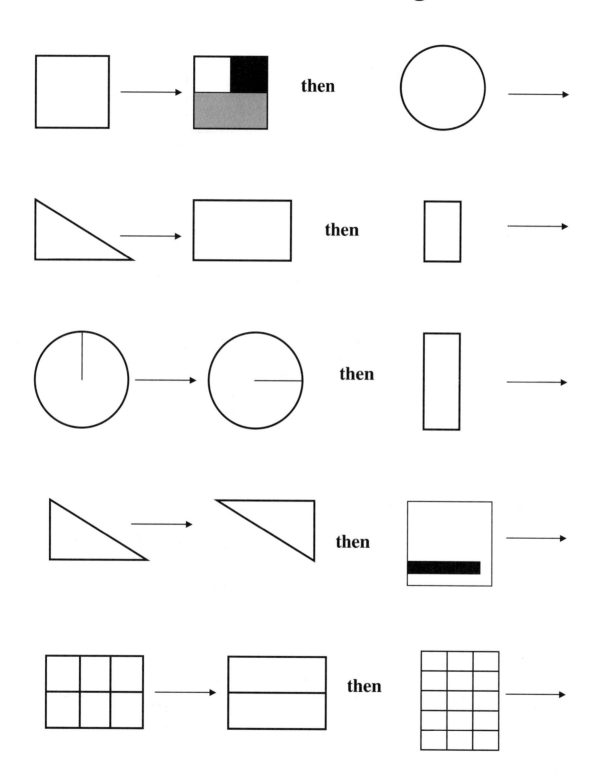

Geometrical Analogies 2

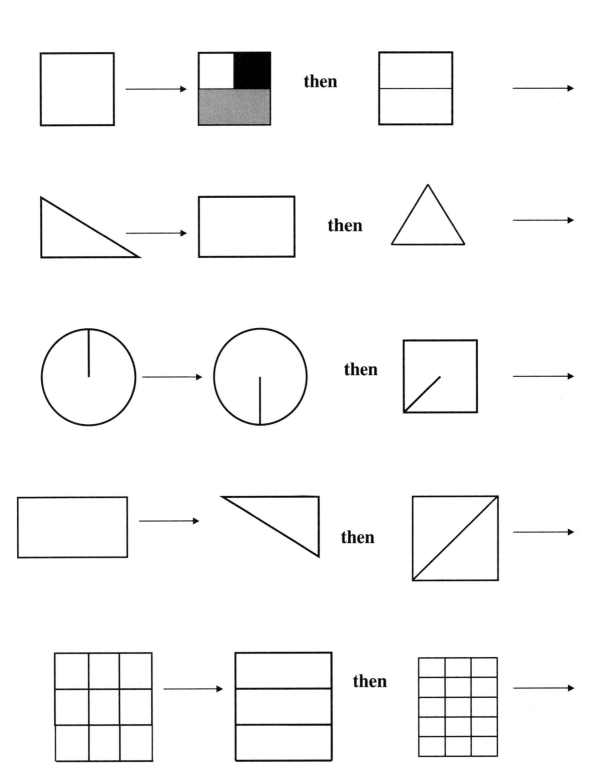

Geometrical Errors

Use the space on the right to describe what is wrong with each figure on the basis of geometrical principles.

a)

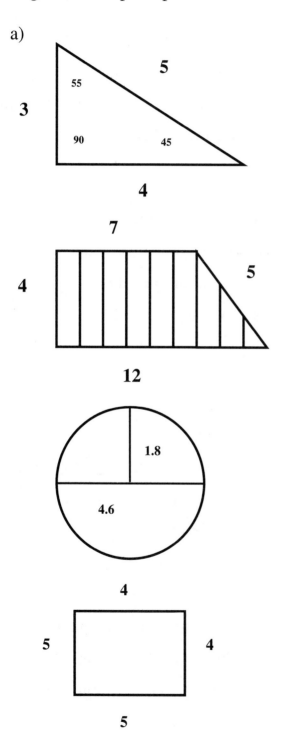

True or False Geo

Circle the statements that are true about each figure

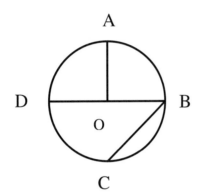

a) There are two pairs of parallel lines
b) There are four visible triangles
c) There are four pairs of congruent triangles
d) $\overline{AB}$ is longer than $\overline{AD}$

e) $\overline{DB}$ is the radius of the circle
f) Three segments shown could be the radius
g) $\overline{CB}$ is a secant of the circle
h) $\overline{AO}$ is ½ of $\overline{DB}$

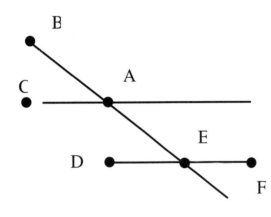

i) m $\angle$ BAC = m $\angle$ AED
j) m $\angle$ BAC + m $\angle$ AEF = 180°
k) BA is perpendicular to DF
l) AEF is complimentary to AED

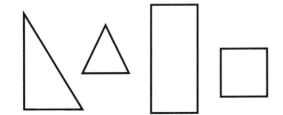

m) All four figures are parallelograms
n) Two figures are rectangles
o) Two figures are congruent
p) There are 9 right angles altogether

Pictorial Mathematics — Geometry

314

Pictorial Mathematics   Geometry

Volume (1)

Find the volume of each shape. Assume each small block measures 1 cm³.

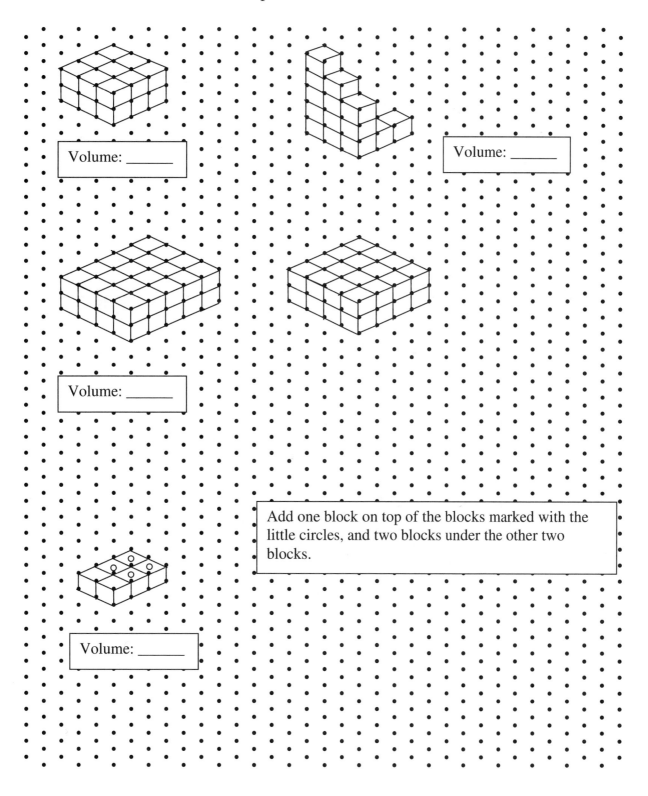

Volume: _____

Volume: _____

Volume: _____

Add one block on top of the blocks marked with the little circles, and two blocks under the other two blocks.

Volume: _____

315

Volume (2)

Find the volume of each shape on the left, then enlarge the figure on the right according to the directions given. Each small block measures 1 cm³.

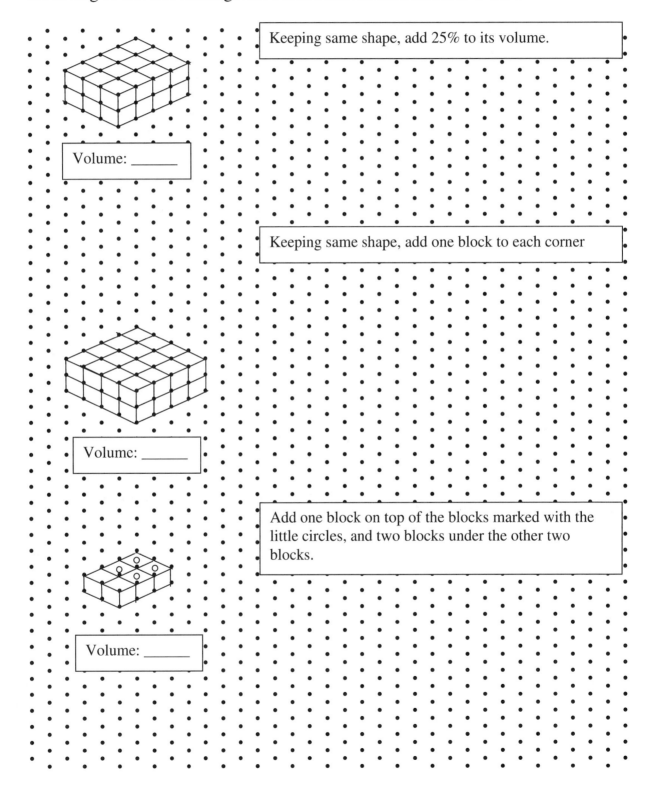

Squares Within Squares

Below are four squares. Each of the smaller squares was made by joining the midpoints of the sides of the larger surrounding square. What percent of the area of the largest square is the shaded square?

Show your work and explain your thinking.

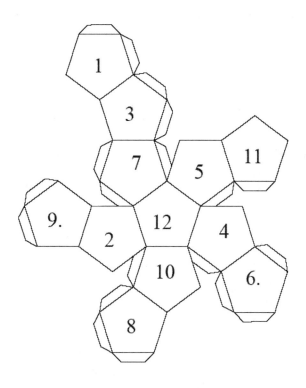

Fold-it

3-Dimensional

Polyhedrons

Teacher Notes
3-Dimentional Folding Polyhedrons

The following few pages contain templates that students can use to build the following eight polyhedrons:

1. Cube
2. Tetrahedron
3. Hexagonal Prism
4. Heptagonal Prism
5. Octagonal Prism
6. Octahedron
7. Cuboctahedron
8. Dodecahedron

The teacher should make copies of the templates onto cardstock of various colors. Students should fold the polyhedrons' faces and glue the corresponding flaps to construct each polyhedron.

Discuss the relevant geometric terms (faces, angles, degrees, edges, vertices, surface area, volume, etc.). An extension project that combines geometry and proportional reasoning is to create templates with the following restrictions:

a) The surface area is twice as large as the original template
b) The volume is twice as large as the original template
c) The edges are twice as long as the original template

After students create the new templates, they should fold them into polyhedrons. Students can also decorate the faces of their polyhedron with photos, theorems, etc. Glue a string to one vertex and hang them from the ceiling to create a great geometric decoration that remind students what they have learned about polyhedrons.

Pictorial Mathematics — Geometry

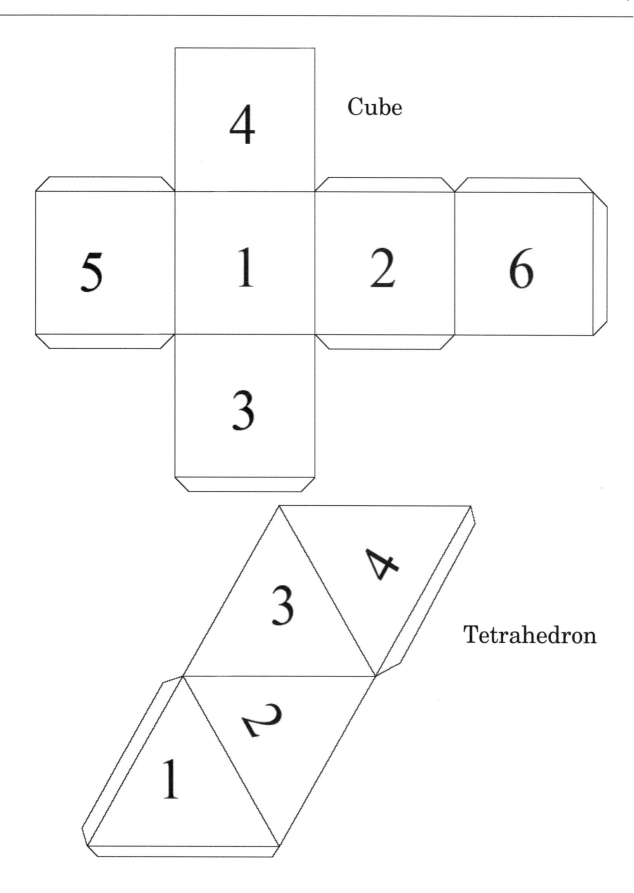

Cube

Tetrahedron

Pictorial Mathematics

Geometry

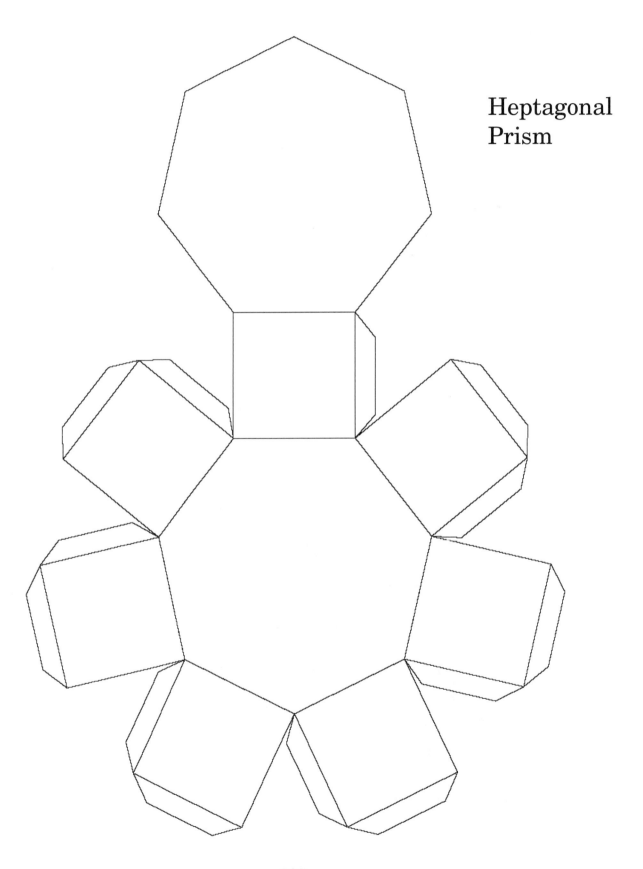

Heptagonal Prism

Pictorial Mathematics — Geometry

Hexagonal Prism

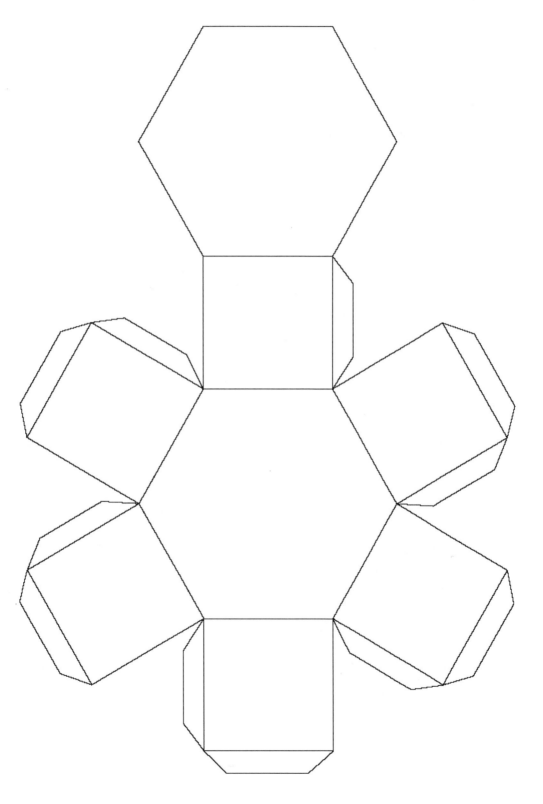

Pictorial Mathematics

Geometry

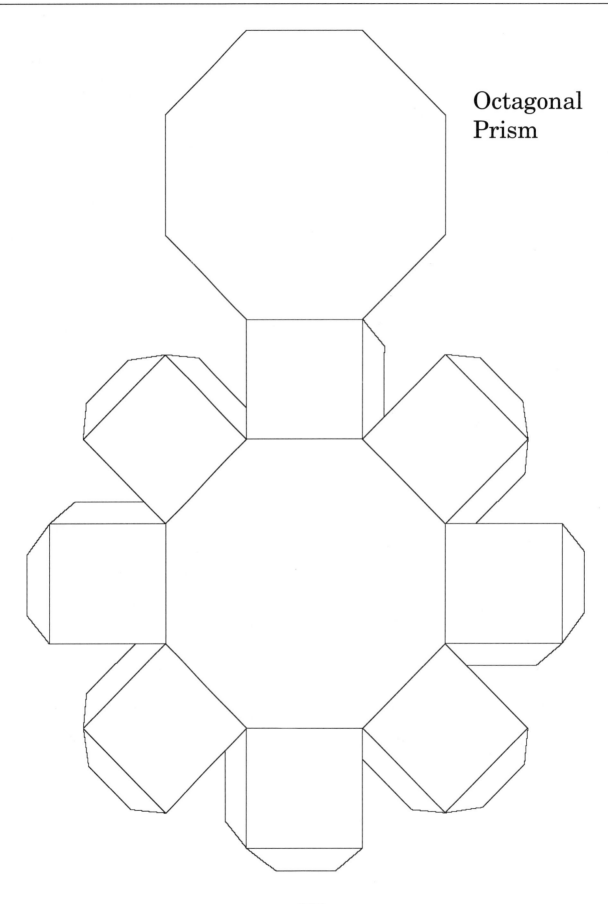

Octagonal Prism

Octahedron

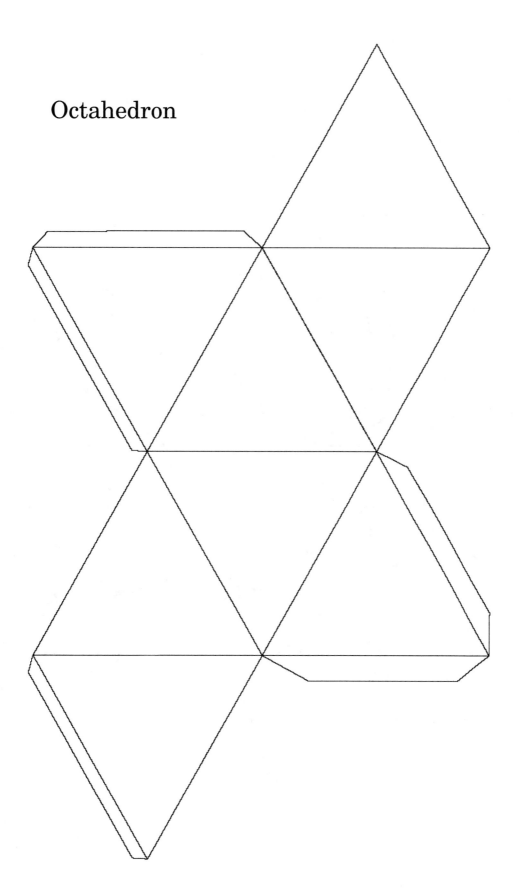

Pictorial Mathematics — Geometry

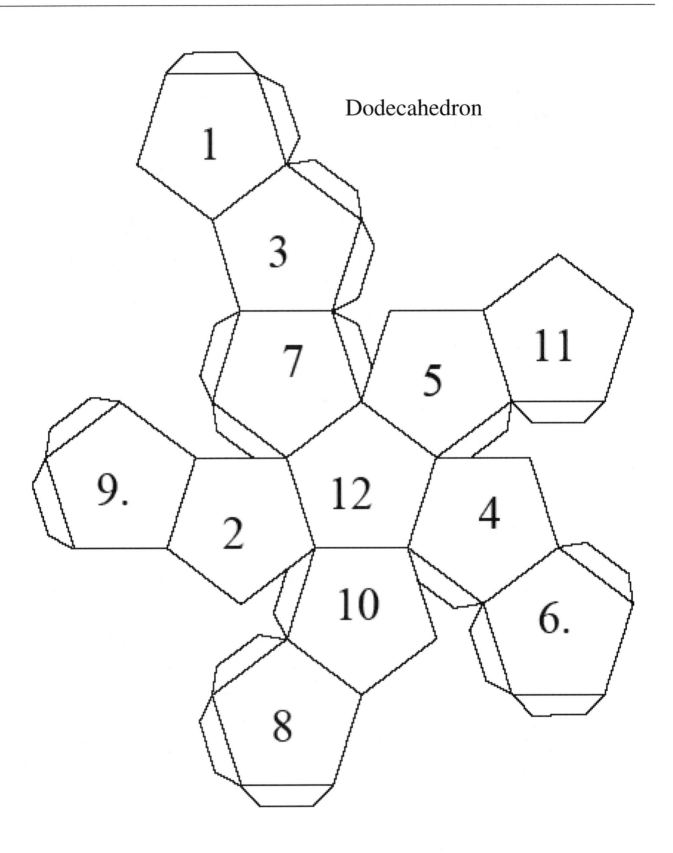

Dodecahedron

Cuboctahedron

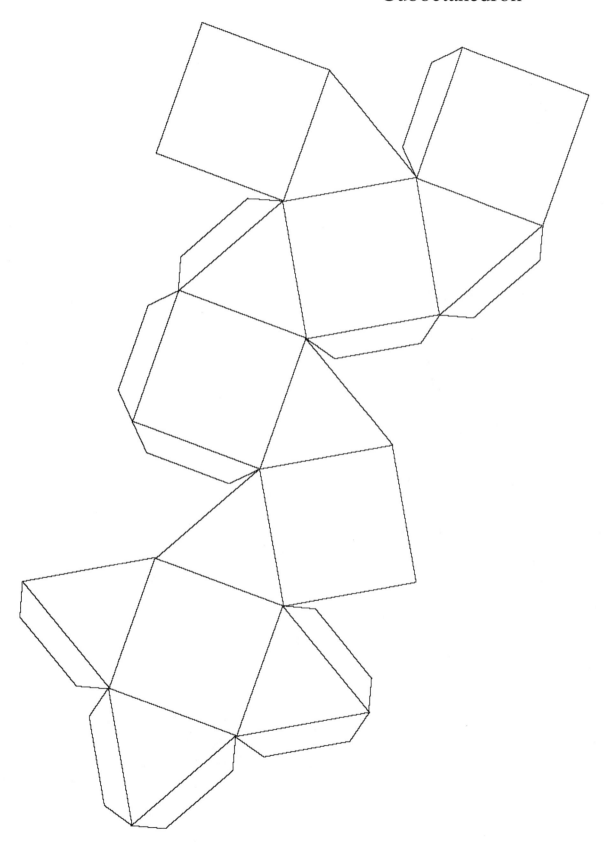

Taking Time To Reflect

Use the grid below and a 1 to 4 scale to estimate how often you ask students to translate concepts from one representational system to another. (4) Very frequently, (3) Frequently, (2) Sometimes, (1) Rarely or never

How Often Do You Engage Students In Translating

From these representations	To these representations					
	Numeric - Algebraic	Pictorial	Concrete - Manipulatives	Oral	Words	Experience Based activity
Numeric - Algebraic						
Pictorial						
Concrete - Manipulatives						
Oral						
Words						
Experience based activity						

The shaded spaces represent the transformations made within each representational system (i.e. numeric to numeric, ½ to 0.5)

Probability and Statistics

Creating a Fair Spinner

You need to create a spinner for a game that gives two players equal chances to get to 40 points. One player will get 1 point every time the spinner lands on an odd number; the other player will get two points every time the spinner lands on an even number. Write the right combination of even and odd numbers inside the blank spinner bellows to make this a fair game. Each slice must have only one number.

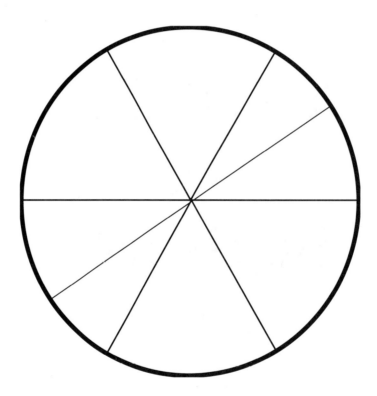

Fair Spinner?

You and a friend are playing a game with the spinner below. One player gets a point every time the spinner lands on an even number; the other player gets a point every time the spinner lands on an odd number. Which player is more likely to get to 10 points first, the one that chooses the even or the odd numbers? Is the game fair? Explain your reasoning.

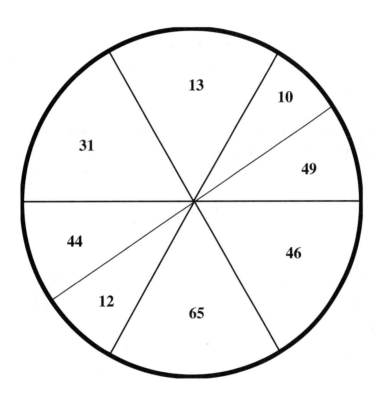

Carnival Spinner

You need to create a spinner for a carnival at your school. Design your spinner with the following rules in mind:

- The chances of the school winning should be two times more than the players' chances; the spinner's slices where the school win should be labeled "School Wins"

- There should be at least one slice where a player would get paid 2 to 1. The chances of landing on this slice should be half the chances of landing on the slices that pay the same amount the player bet.

- The number of slices and their position should make players believe they have a reasonable chance to win.

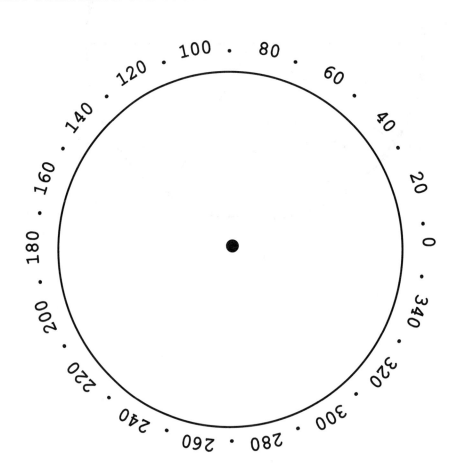

Dart's Chances

What is the probability of landing in each of the four rings of the dart board below?

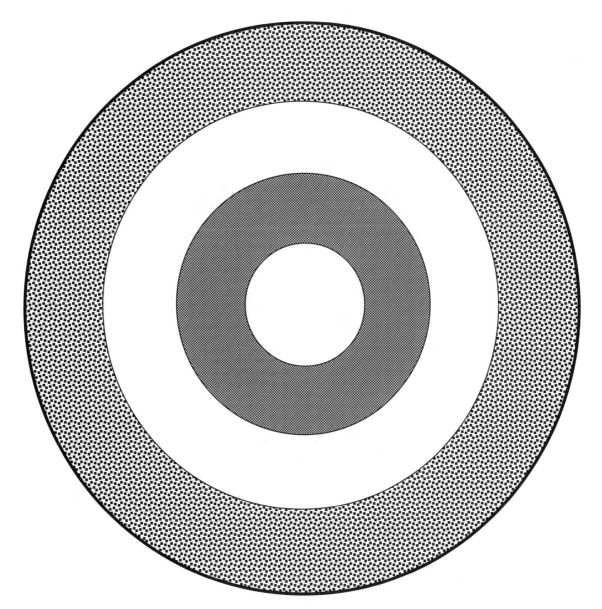

Challenge question: Suppose you get 1 point for landing in the outermost ring, 2 points for the next ring, 3 points for the next, and four points for landing in the innermost ring. Consider the probabilities of landing on each ring. If you want to be the first to get to 12 points by hitting the same target with each shot, where should you aim? (you would only get points for landing on the ring selected at the start of the game.)

"*If you think education is expensive, try ignorance*". ~Andy McIntyre

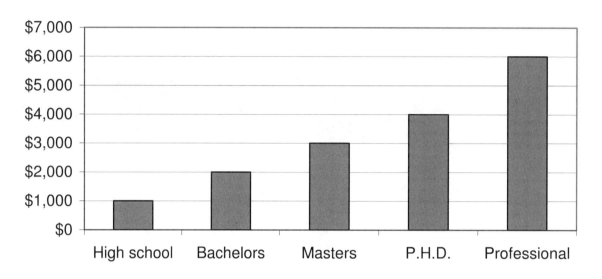

Create a pie graph from the data given in the bar graph above. The total of all the pie pieces should equal the combined income by all five groups above. Use different shading patterns for each group.

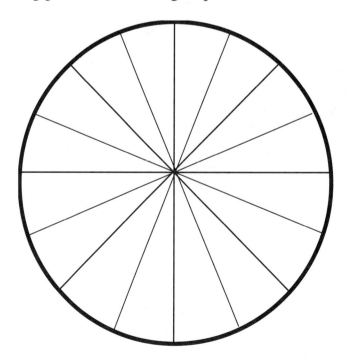

Pictorial Mathematics · Probability and Statistics

Graphing Transformations

Keeping the same proportions, create the graph on the right from the data on the left.

Bar graph

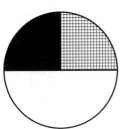
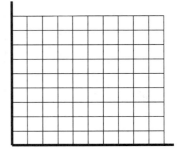

Bar graph

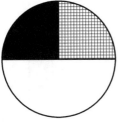
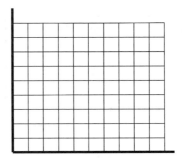

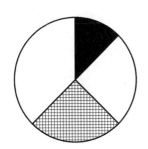

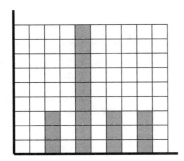
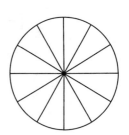

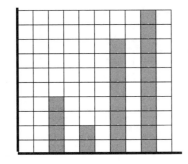

335

How Many Students Brought Their Homework? (1)

Chart and graph the number of boys and girls who bring their homework on time each day for two weeks.

Days of The Week	Mon	Tue	Wed	Th	Fri	Mon	Tue	Wed	Th	Fri
Number of Students										

B	G	B	G	B	G	B	G	B	G	B	G	B	G	B	G	B	G	B	G
Mon		Tue.		Wed.		Thur		Mon		Tue		Wed		Thur	

How Many Students Brought Their Homework? (2)

Chart and graph the number of boys and girls who bring their homework on time each day for two weeks. Use the data to find the range, mean, median and mode.

Days of The Week	Mon	Tue	Wed	Th	Fri	Mon	Tue	Wed	Th	Fri
Number of Students										

Range	Mean	Median	Mode for boys	Mode for girls

How high does the ball bounce?

Using three different types of bouncing balls, find out how high each ball bounces as you drop them from five different heights. Measure and chart the height of the bounce twice for each drop. Graph the data and use it to predict how high the ball will bounce from the five heights given below.

Height of Drop									
Height of Bounce									

Predict how high each ball will bounce when dropped from each of the heights given below

Height of Drop	_____	_____	_____	_____	_____
Ball A					
Ball B					
Ball C					

Birthday Graph (1)

Graph the number of birth dates of boys and girls by month. Label your graph, and use different colors for boys and girls.

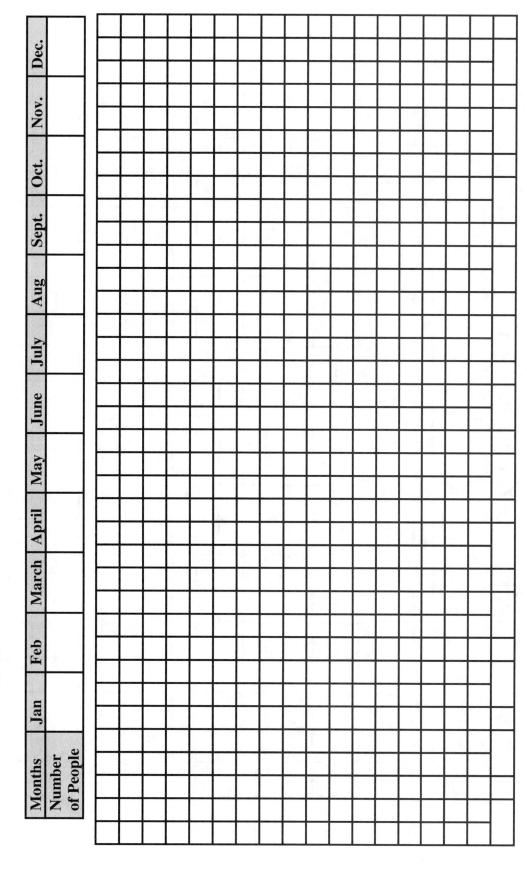

Birthday Graph (2)

Chart the number of birth dates of male and female students by month. Graph the data and use this information to find the range, mean, median and mode.

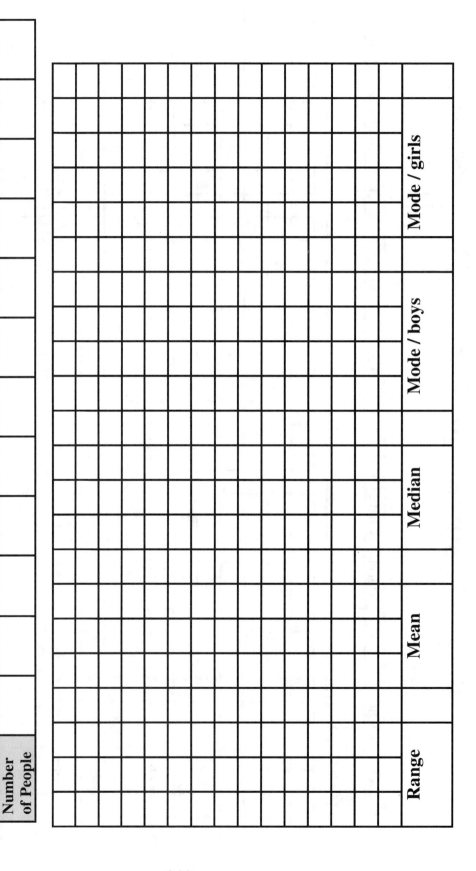

Counting and Eating M&M's

	Prediction	Actual
How many M&M's are in your bag?		
What is the most common color?		
What is the least common color?		
What is the class favorite color?		
What is the class least favorite color?		

Individual Data **Data for your small group**

Colors	Total #	Fraction	Percent	Total #	Fraction	Percent

Graph your results

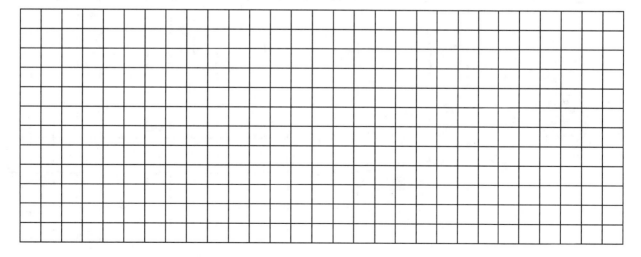

How large is your shoe?

Each group measures and reports their shoe lengths to the class. Each student will use the first table to create a frequency table. Use the grid below to graph your results. What is the average shoe length? What is the mode?

Length Range	Number of Shoes

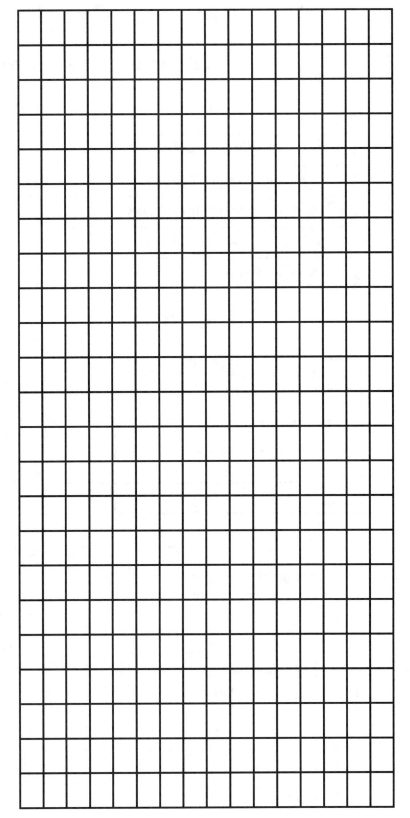

Pictorial Templates

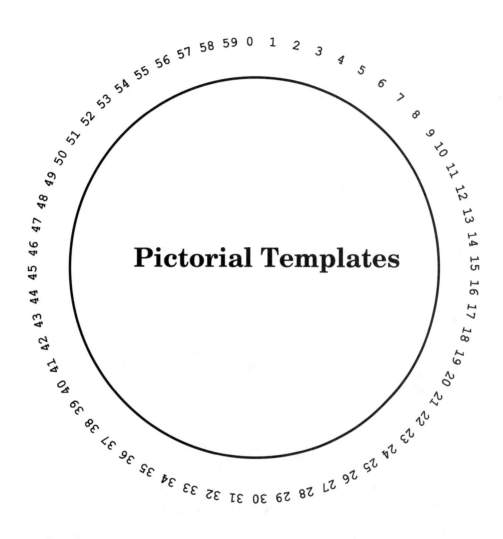

Pictorial Templates

This section contains 38 graphic templates that can be used to teach, practice and assess just about every topic in this book. At the end of the section I have included eighty sample ideas on how to use these templates to engage students in learning about:

- Number representations
- Equivalent fractions
- Graphing equations
- Graphing inequalities
- Data analysis
- Decimals
- Percents
- Adding and multiplying fractions
- Ratios
- Proportions
- Area
- Perimeter
- Multiplication
- Geometric concepts

I strongly recommend that teachers make transparencies of the pictorial templates that students will use. Use these transparencies to model how to effectively use the template with a few examples.

Instructional Tips:

- Laminate the pictorial templates you might use most.
- Whenever appropriate, ask students to explain their reasoning.
- Ask students to show alternative solutions.
- Whenever appropriate, ask students to generalize their findings.
- Vary the way students complete the tasks, individually, in partners, or in small groups.
- Whenever possible, give student access to the appropriate manipulatives, such as base-10 blocks or algebra tiles.
- In your planning, do the problems yourself prior to giving them to your students.
- Take a few moments before teaching to write down the specific language you want students to associate with the pictorial models (i.e. 2 repeated three times).

Pictorial Mathematics　　　　　　　　　　　　　　　　　　　　　　　　　　Appendix

10 by 10 Grids

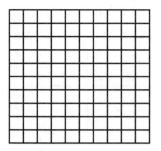 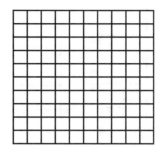 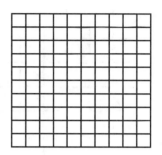

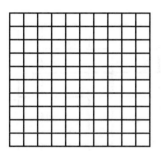 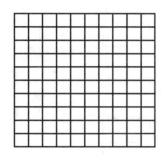 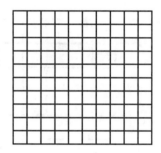

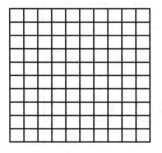 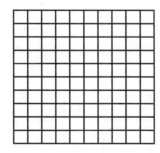 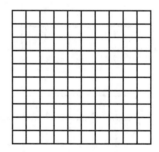

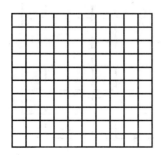 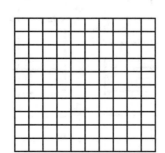 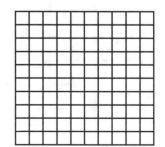

Pictorial Mathematics — Appendix

Six 10x10 Grids

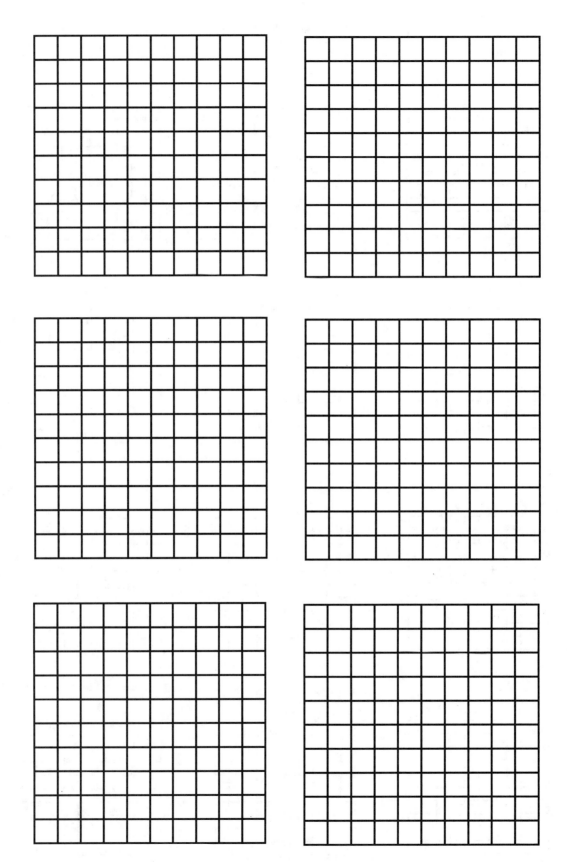

10 by 10 Grid

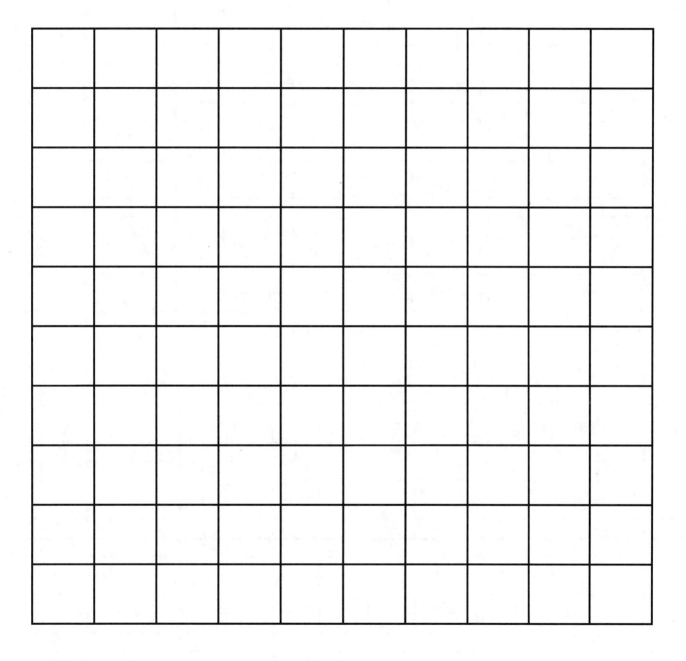

Graphing Grids

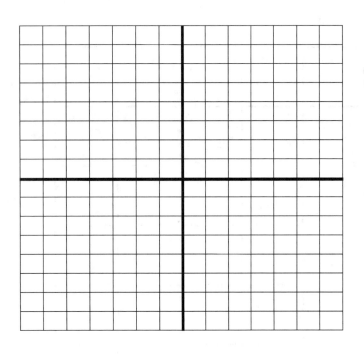
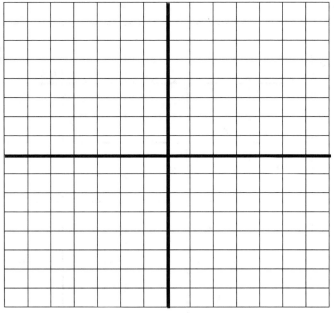
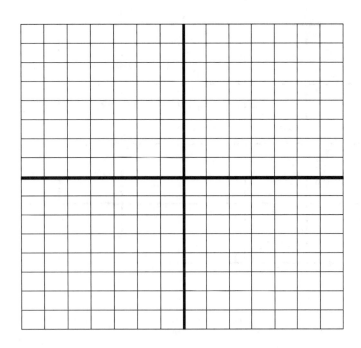
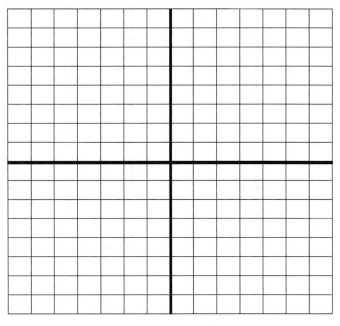

1-Inch Graph Paper

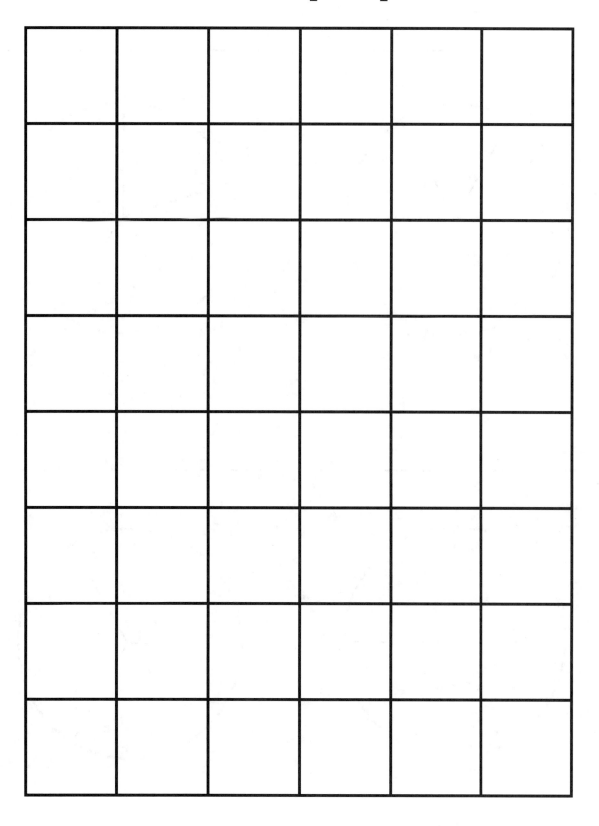

Circle Fractions 1

_____   _____

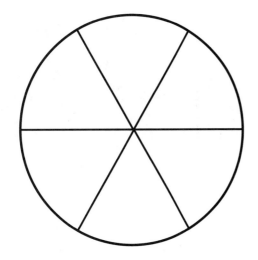   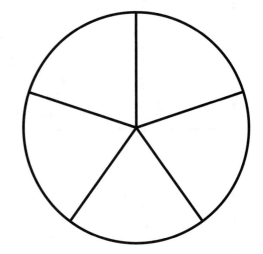

_____   _____

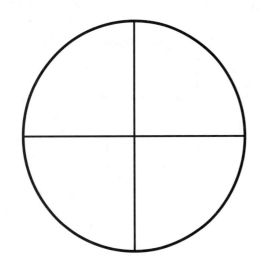   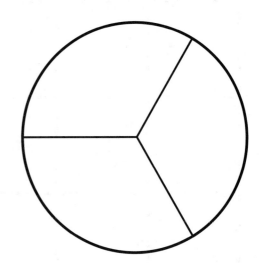

Circle Fractions 2

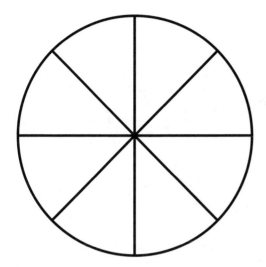

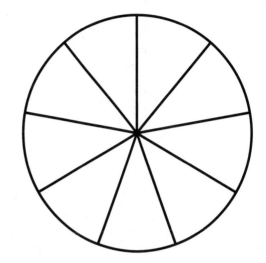

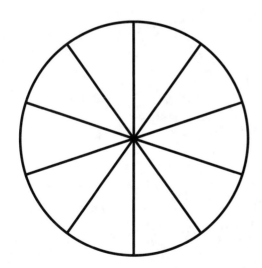

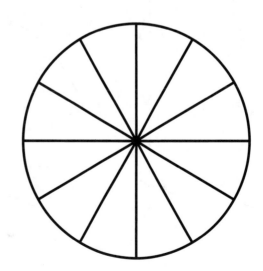

Pictorial Mathematics   Appendix

Circle Fractions 3

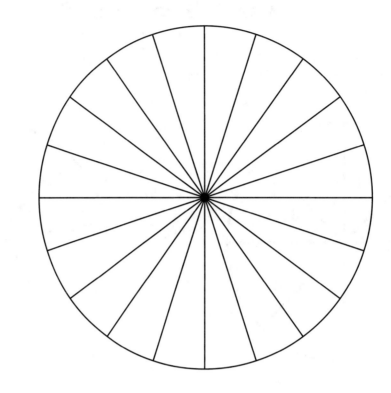

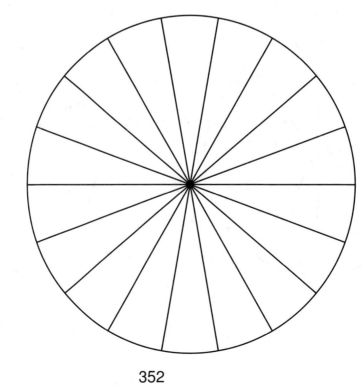

Pictorial Mathematics   Appendix

24 hour Clock

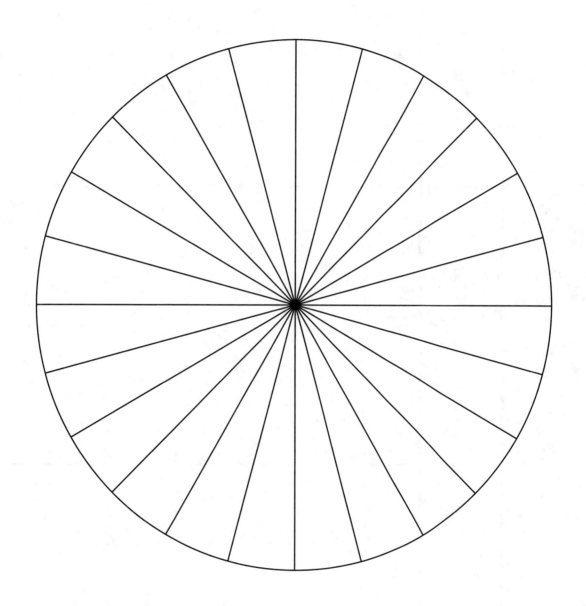

Whole Number Place Value Mat

Hundreds □	Tens ▭	Ones ▫

Pictorial Mathematics — Appendix

Decimal Place Value Mat

Ones □	Tenths ▭	Hundredths ▫

Write the numbers constructed above, in sequence and in standard form, on this space.

Base-10 Manipulatives

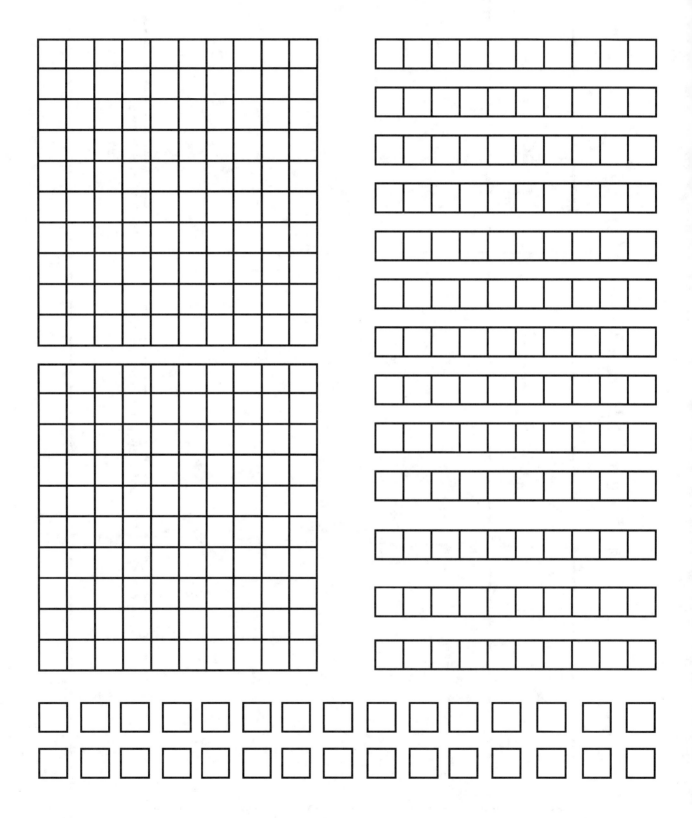

Geoshapes

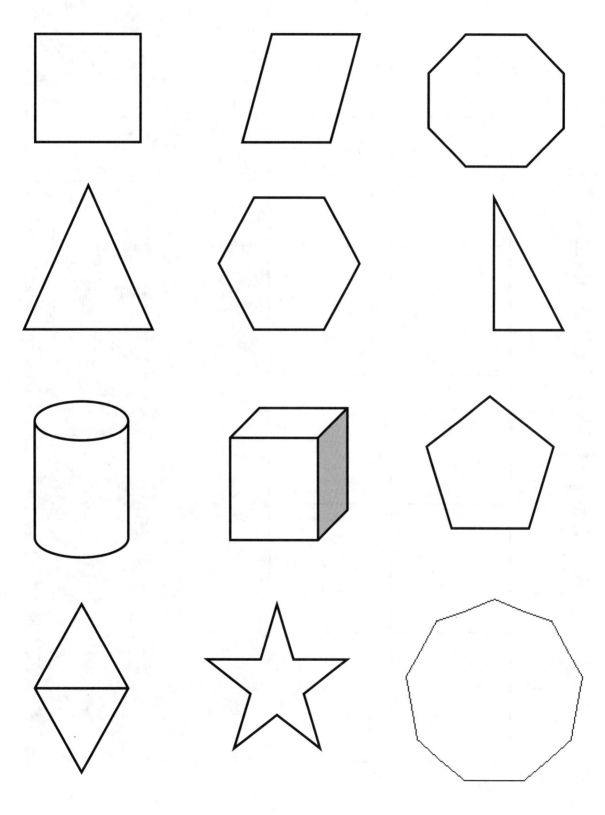

Pictorial Mathematics Appendix

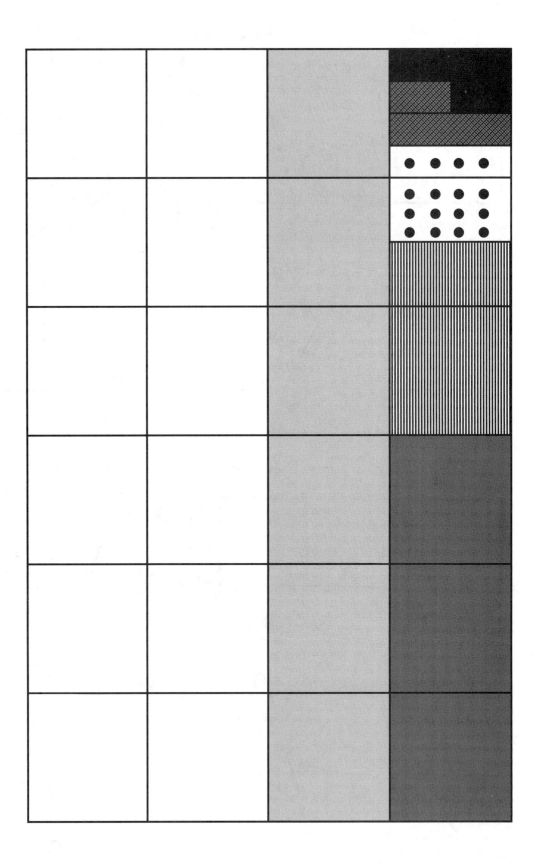

Large 6 by 4 Half of Half Grid

Pictorial Mathematics Appendix

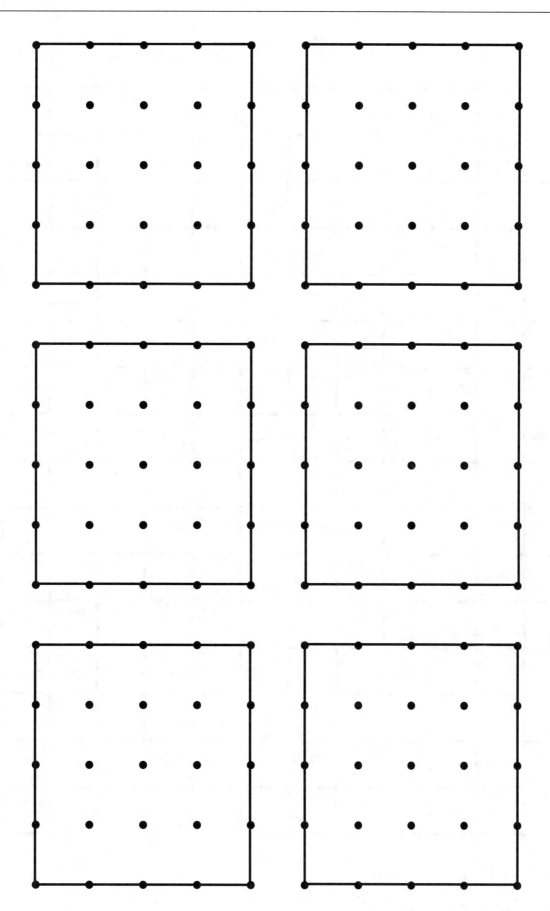

4 by 4 Square Dot Paper

Pictorial Mathematics — Appendix

Half-Inch Graph Paper

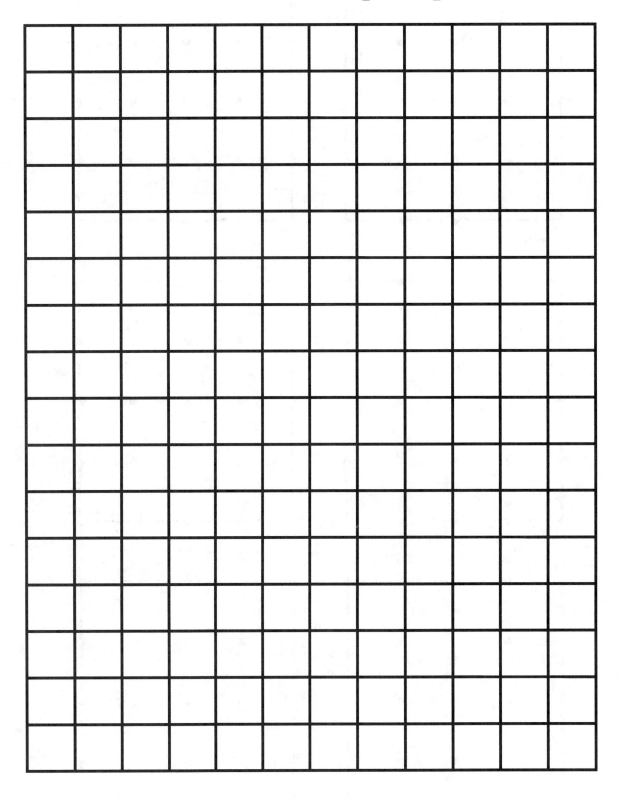

Quarter Inch Square Paper

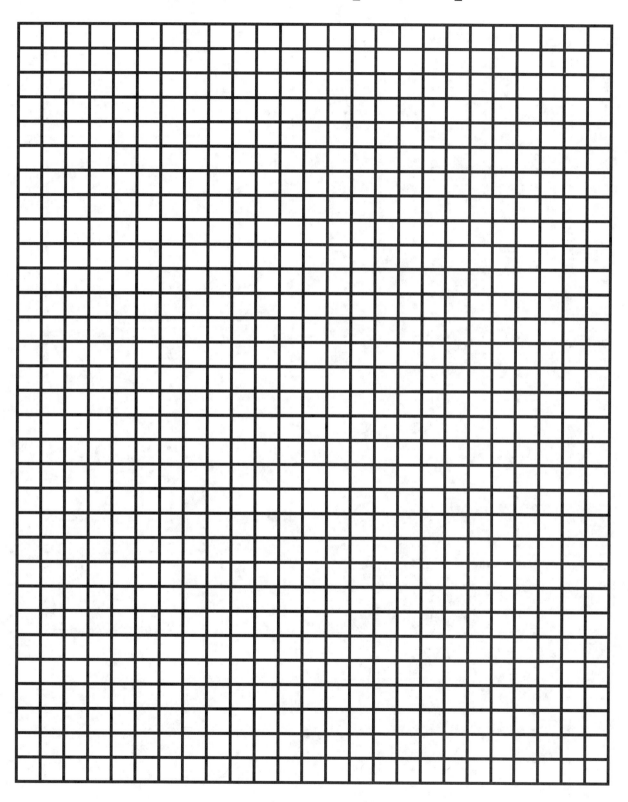

Isometric Paper

Pictorial Mathematics — Appendix

Isometric Squares

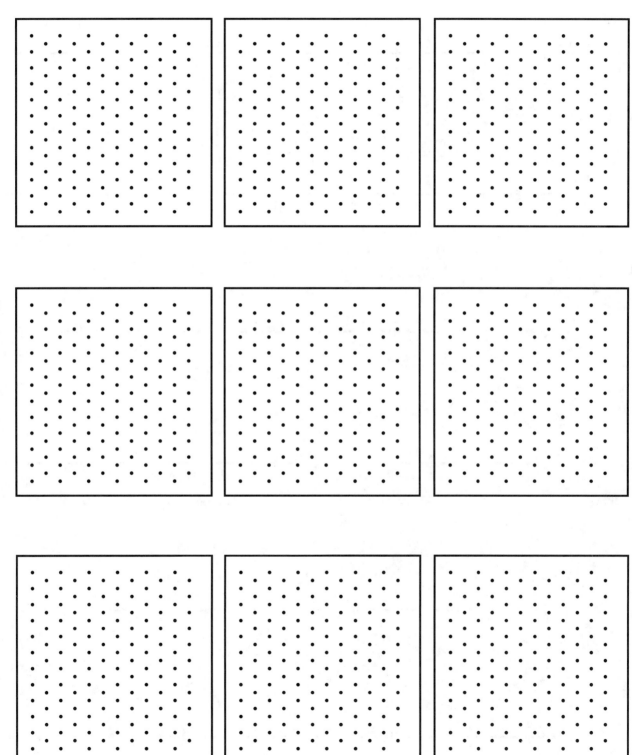

363

Number Lines

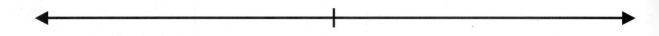

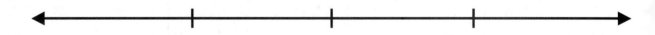

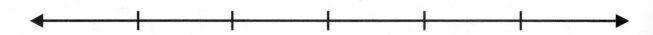

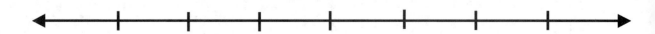

Pictorial Worksheets

4 by 3 Grid

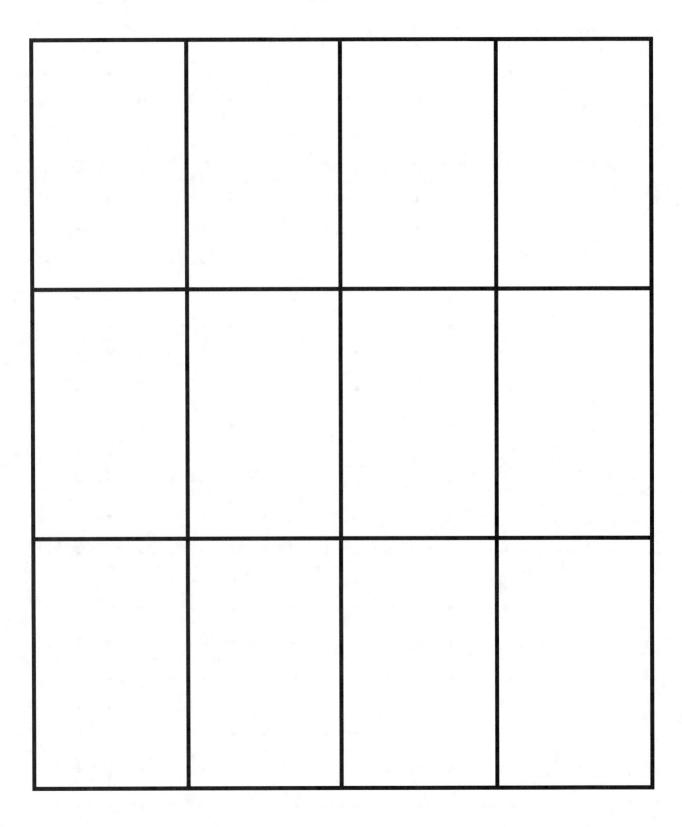

4 by 4 Grid

3 by 2 Grid

100's Chart

1	2	3	4	5	6	7	8	8	10
11	12	13	14	15	16	17	18	19	20
21	22	23	24	25	26	27	28	29	30
31	32	33	34	35	36	37	38	39	40
41	42	43	44	45	46	47	48	49	50
51	52	53	54	55	56	57	58	59	60
61	62	63	64	65	66	67	68	69	70
71	72	73	74	75	76	77	78	79	80
81	82	83	84	85	86	87	88	89	90
91	92	93	94	95	96	97	98	99	100

Pictorial Mathematics — Appendix

Hundredths and Tens Charts

.01	.02	.03	.04	.05	.06	.07	.08	.09	.10
.11	.12	.13	.14	.15	.16	.17	.18	.19	.20
.21	.22	.23	.24	.25	.26	.27	.28	.29	.30
.31	.32	.33	.34	.35	.36	.37	.38	.39	.40
.41	.42	.43	.44	.45	.46	.47	.48	.49.	.50
.51	.52	.53	.54	.55	.56	.57	.58	.59	.60
.61	.62	.63	.64	.65	.66	.67	.68	.69	.70
.71	.72	.73	.74	.75	.76	.77	.78	.79	.80
.81	.82	.83	.84	.85	.86	.87	.88	.89	.90
.91	.92	.93	.94	.95	.96	.97	.98	.99	1.0

10	20	30	40	50	60	70	80	90	100
110	120	130	140	150	160	170	180	190	200
210	220	230	240	250	260	270	280	290	300
310	320	330	340	350	360	370	380	390	400
410	420	430	440	450	460	470	480	490	500
510	520	530	540	550	560	570	580	590	600
610	620	630	640	650	660	670	680	690	700
710	720	730	740	750	760	770	780	790	800
810	820	830	840	850	860	870	880	890	900
910	920	930	940	950	960	970	980	990	1000

Bar Graph Grids

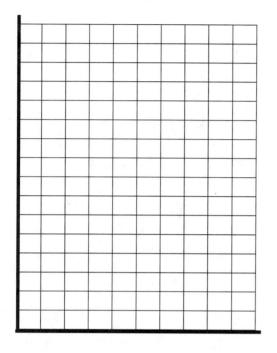
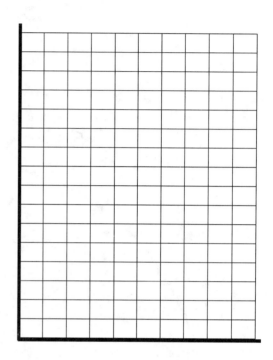
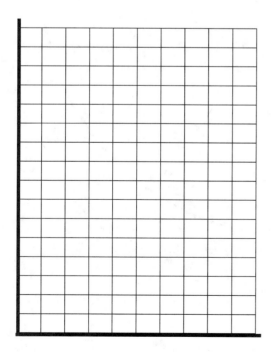
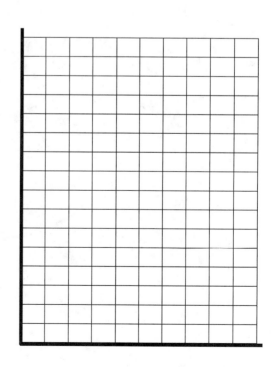

Pictorial Mathematics Appendix

60 Second Clock

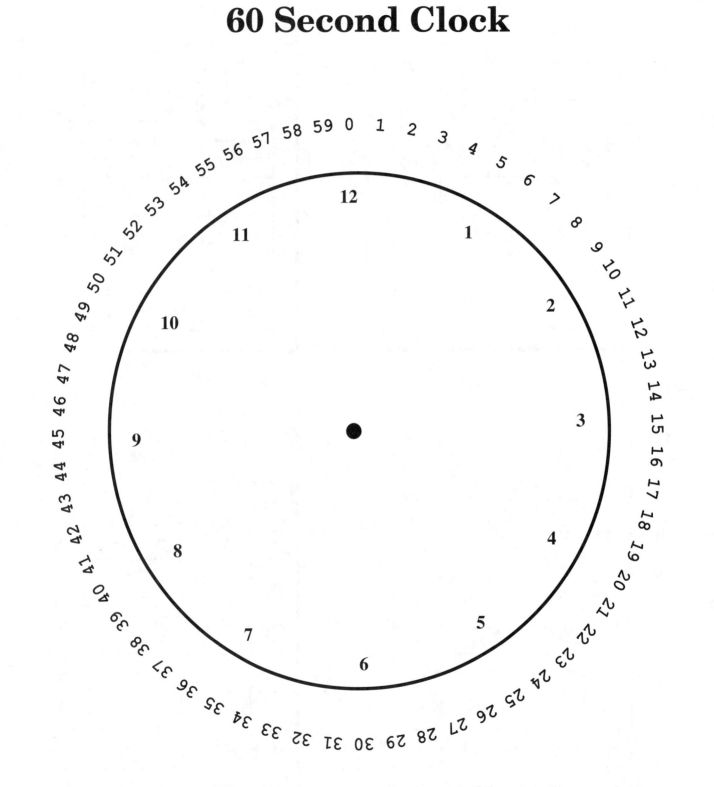

Pictorial Mathematics   Appendix

Geoboard Paper

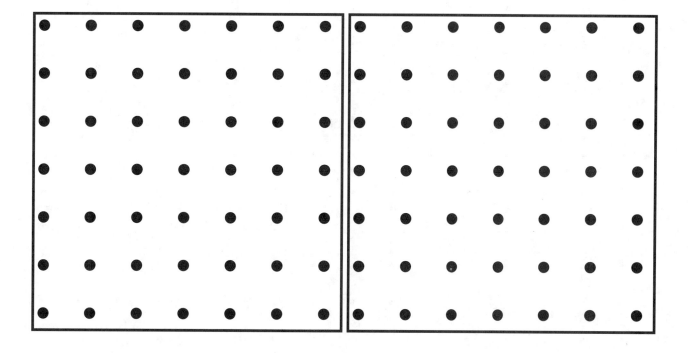
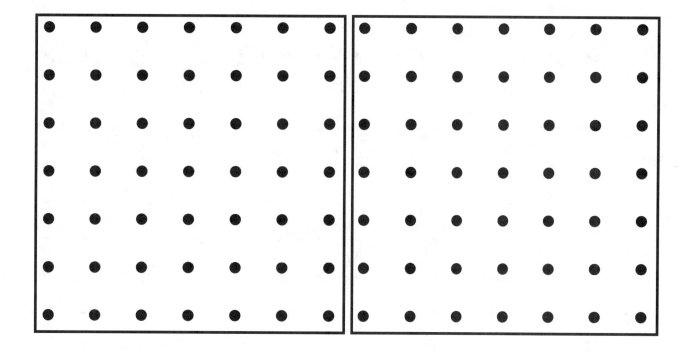

Dot Picture Paper

Pictorial Mathematics Appendix

Twelve Circles

375

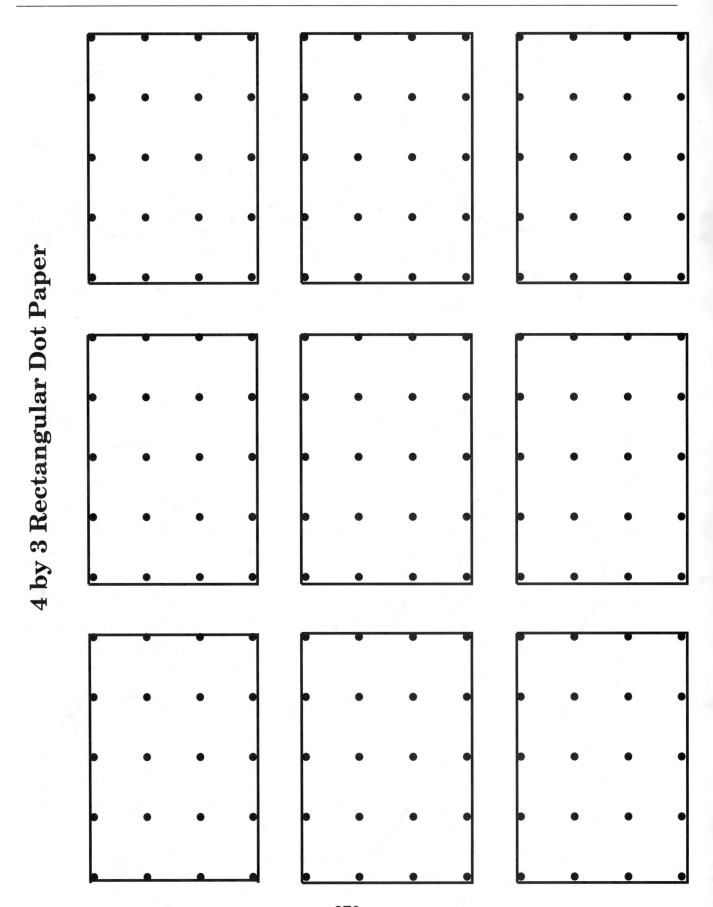

Pictorial Mathematics

Large 4 by 3 Grid

Large 6 by 4 Grid

Tables 1

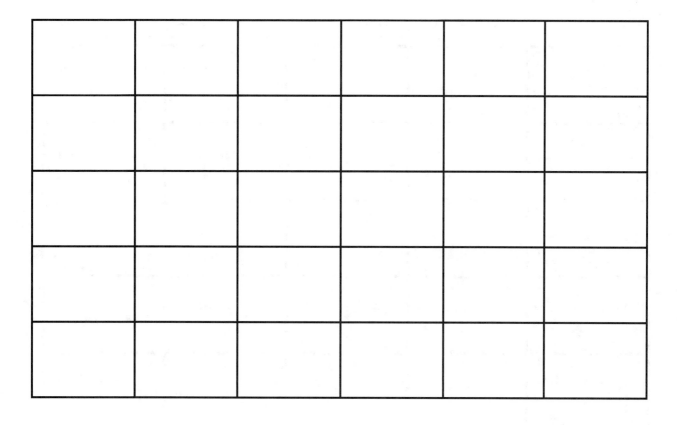

Tables 2

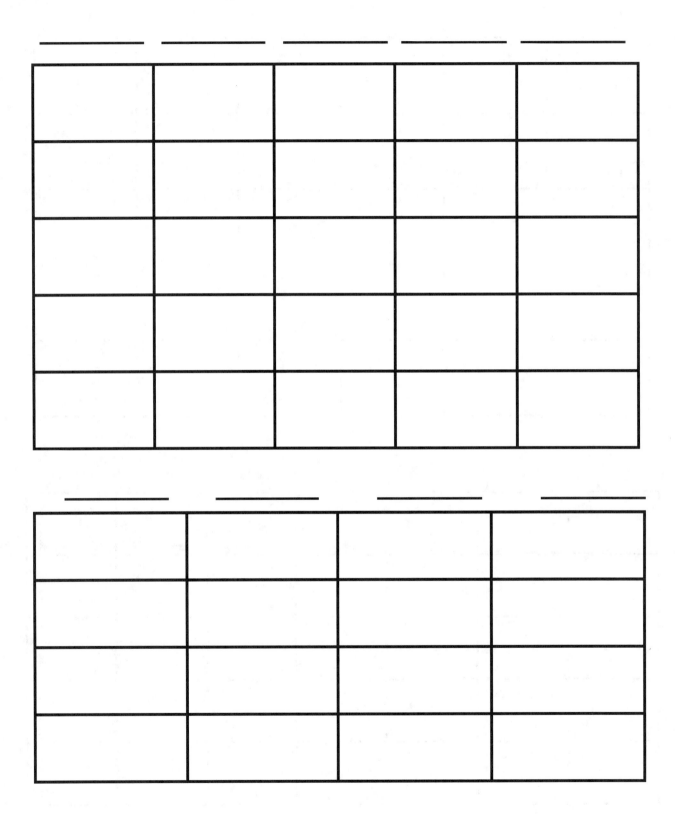

Pictorial Mathematics — Appendix

Tables 3

Tables 4

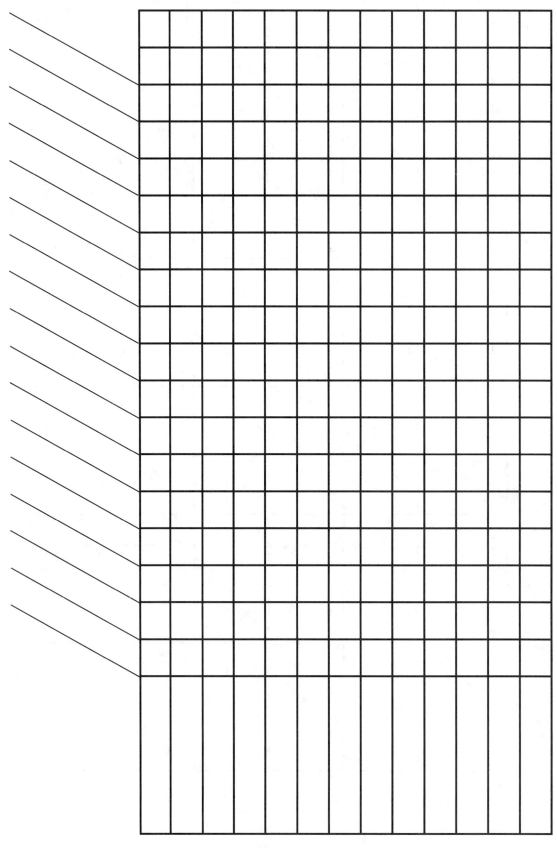

Pictorial Mathematics | Appendix

Teacher Notes – Pictorial Templates

Sample Activities/Tasks Based On Pictorial Templates

This section contains eighty sample exercises and activities that illustrate how teachers might use the pictorial templates to further develop conceptual understanding, give students more skill practice, or engage them in problem solving.

I strongly recommend that teachers and parents who use these templates to create additional practice problems for their students do the problems they create for themselves first, to ensure that they have created workable, worthwhile problems.

10 by 10 Grids (p. 345-47)

1) Shade the grid to pictorially represent standard numerals in many forms, such as: 0.04, $\frac{7}{10}$, $1\frac{3}{4}$, 69 %, etc.

2) Shade the grid to pictorially represent addition of standard numerals such as: $\frac{3}{4} + \frac{1}{2}$, $1\frac{3}{8} + 0.4$. etc. Be sure to change the representation of the numerals (fractions, decimals, whole numbers, etc.),

3) Shade-in percentages, such as 45%, 12.5%, 40% of the first row, etc.

4) Create a daily bar graph that charts the changes in the daily prices of three or more stocks.

5) Trace small objects of various shapes over onto the grid. Have students find the objects' areas/perimeters and explain the process used.

6) Draw/outline all possible rectangles with an area of 12, 25, 30, and 40 square units. For example, all the following rectangles have an area of 12 square units: 12 by 1, 6 by 2, and 4 by 3.

7) Outline rectangles and use different colors to shade-in and name various fractional parts of the rectangle.

8) Estimate the size of square grids with 10,000 and 1,000,000 squares knowing that the 10-by10 grids are made of 100 squares.

9) Figure the smallest perimeter of a rectangle made of 1,000 of the small grid squares. Repeat for 1,000, 2,000, 3,000 …to 10,000 squares. Look for and describe any patterns observed.

10) Using one grid for each rectangle, outline rectangles with areas of 12 squares, 24 squares, 36 squares and so on through several multiples of 12. Shade each grid so that the ratio of shaded to un-shaded squares in all the grids is 1:3, 1:5, 2:1, or 11 to 1.

Graphing Grids (p. 348)

11) Plot the following points (0, -1), (-3, 0), (4, 1), then find the slope of each of the segments connecting the points.

12) Graph sets of three equations on single grids. For example:

>   (grid 1): $y = x$, $y = x + 1$, $y = x - 1$
>   (grid 2): $y = x$, $y = 2x$, $y = -2x$
>   (grid 3): $y = x$, $y = \frac{1}{2}x$, $y = -\frac{1}{2}x$
>   (grid 4): $y = x$, $y = 2x + 1$, $y = 2x - 1$

Describe the patterns and changes observed within and across grids.

13) Graph one equation, such as $y = 2x$, on a grid. Graph the same line but moved horizontally and vertically a certain number of units. Write the equation of the new lines.

14) Graph the following inequalities:

>   (grid 1): $y > x$
>   (grid 2): $x > y$
>   (grid 3): $y > x + 1$

(grid 4): $y < x - 1$

Describe the difference and similarities across the four graphs.

15) In pairs, take turns drawing a line on a grid without letting one's partner watch. The drawer should describe the line in as few words as possible so that his or her partner can draw the exact same line on his or her own grid. To make the activity more challenging, do not allow students to use certain words or terms, like specific coordinates (do not say "one, minus 2" for example.)

1-Inch Graph Paper (p. 349)

16) Create bar graphs from data collected, from a chart, from a line graph, from a pie chart, etc.

17) Outline a 4 by 6 rectangle with a marker or crayon. Color $\frac{1}{2}$ of the rectangle yellow, $\frac{1}{4}$ of the rectangle blue, $\frac{1}{8}$ of the rectangle brown, and $\frac{1}{16}$ of the rectangle green.

18) Each row of the grid has seven squares, one for each day of the week. There are eight rows of seven columns, representing a total of 8 weeks. What fraction of the days on the grid are weekend days? Keep track of the total number of school weeks and days left in the school year. Ask questions such as: What percentage of the days left in school are Fridays?

19) During the first 2 minutes of class, write a basic fact-family in the first square, for example: the two's multiplication tables (2X). Keep writing the same fact-family in successive squares until the facts are memorized.

20) Outline a 4-by-6 rectangle. How many rectangles can be counted within this rectangle? How many squares? Repeat this exercise with a 6-by-6 and a 6-by-8 rectangle. Is there a pattern? If so, describe it.

Pictorial Mathematics — Appendix

21) Cut out the margin around the grid so that only the 7 by 8 grid remains. Figure out how to cut seams and fold the grid to make a box that can hold 18 cubic inches.

22) In groups of three or four, construct all the different boxes that can hold 24 cubic inches.

23) In groups of three or four, determine the largest volume rectangular prisms and cubes that can be constructed with one and two grids.

24) Find he box with the smallest surface area that can hold 12 cubic inches. Repeat for surface areas of 16, 18 and 24 cubic inches.

25) With red, yellow and blue crayons or markers, color a 6-by-4 rectangle so that the ratio of red to blue is 1:2, and the ratio of blue to yellow is 2:3. Extend this line of questionings by changing the size of the rectangle, the number of colors and the ratios involved.

4 by 3 Rectangle Dot Paper (p. 376)

26) Divide and shade three rectangles so that the sum of the shaded parts of the first two rectangles equals the shaded fraction of the third rectangle.

27) In groups of three or four, divide and shade the rectangles to show the following: $(1\frac{1}{2}+\frac{4}{3})$, $(\frac{1}{8}+\frac{1}{4}+\frac{5}{6})$, and $(\frac{5}{12}+\frac{5}{4}+\frac{5}{6})$. Cut out the rectangles and glue them to a piece of construction paper. Show at least two different ways to construct the required combinations.

28) In pairs, take turns drawing a shape of one's choosing without the partner watching. Attempt to describe this shape to the partner so that the partner can duplicate the shape.

29) Using all rectangles, create ratios of shaded to un-shaded rectangles, such as: 1 : 11, 1 : 5, 1 : 3, 1 : 2, 5 : 7, 1 : 1, 7 : 5, 2 : 1, and 3 : 1.

Circle Fractions 1, 2 and 3 (p. 350-52)

30) Create fair spinners.

31) Shade-in each circle to show a given fraction.

32) Shade a given percentage red, and a given fraction blue.

33) In groups of three to five, cut out the circles. Create a poster board showing all possible equivalent fractions.

34) Cut out the circles. Show which circles could not be used to show a given fraction.

35) Select an appropriate circle and use different colors to shade in the monetary breakdown of various scenarios, for example: the total dollars spent for a party is $360 dollars, of that, $60 was spent on decorations, $90 on desserts, $120 on food, and the rest on drinks.

36) Each week, choose and color the appropriate circle to show the percentage of students who submit homework on time, who wear red t-shirts, etc.

37) Create a pie chart from a given line graph, bar graph, chart, or from data collected in survey.

38) Create a fair spinner that would pay 2 to 1 if it lands in yellow, 3 to one if it lands in red, and 1 to one if it lands in blue. Spin the spinner thirty times and record the results on a chart. How did the actual spinning results compare with the spinner's theoretical probability of landing on each section?

24-Hour Clock (p. 353)

39) Create a pie-chart that shows how much time you spend: a) at school, b) eating, c) sleeping, d) playing, e) doing homework, f) doing other things

40) Write the time of the day corresponding to each slice of the circle. Answer questions such as: what is the time if 1/6 of the day has gone by? What fraction of the day has passed by 3:00 p.m.?

Place Value Mats (p. 354-55)

41) Construct given written numbers with the manipulatives on top of the mat.

42) For each number called out by the teacher, construct it with the manipulatives and write the number on the space provided at the bottom of the mat.

43) After constructing a number on the mat, add or to subtract a given amount from that number; i.e. subtract 3 tens, subtract 26, add six ones.

44) Pair up. Have one student do exercises 41-43, while the partner draws the numbers being constructed using the pictorial representations defined in page 43.

45) Still in pairs, have one student add and subtract numbers with the aid of the mat; the partner adds and subtracts the standard way. Take turns using the mat.

Base-10 Manipulatives (p. 356)

This template is meant to be used whenever actual base-10 manipulatives are not available. Copy the Base-10 manipulative template onto card stock and have students cut the pieces out and put them in envelopes with their names. Paper manipulatives are easy to replace and students can create their own-take home set to practice a wide variety of activities dealing with place value, multiplication, addition, subtraction, division, fractions, percents, perimeter, area, etc.

47) Students use the manipulatives to construct a variety of numbers written on the board by the teacher. Start with whole numbers.

48) Students use the manipulatives to construct a variety of numbers read by the teacher.

49) In pairs, then individually, construct decimals with the manipulatives.

50) Using the base-10 double-digit multiplication examples in page 74, construct double-digit multiplications with the base-10 strips. After constructing the given multiplication, write the results in expanded notation.

GeoShapes (p. 357)

51) Over one week, write everything you know about the geometric properties of each shape. This should include, when appropriate: angles, perimeter, area, volume, symmetry lines, number of parallel sides, surface area, etc. As a class, post three to five new ideas or terms for each figure each day on the bulletin board.

52) Using the dot paper template along with the geoshape template, enlarge each shape by a given factor. For example: make their area twice as large, increase their perimeter 1.5 times.

Large 6 by 4 Half of Half Grid (p. 358)

52) Label each shaded section as a, b, c, etc. Identify what fractional part of the whole grid is a, b, c, etc. For example, section (a) is ½ of the whole grid.

53) Ask students what fraction of the whole grid would equal a+c, d+e, d+e+f, a-e, etc.

54) Assign a dollar value to the whole grid, such as $6, or $12. Find the value of a, b, c+a, etc.

55) Assign a dollar value to specific sections, then find the value of the other sections.

56) In groups of four, place the grid on the floor and frame its border with math books. Only the grid should be visible from top. Drop a small ball of paper onto the grid 24 times. First, find the percentage of time that the ball lands on each section. Then, find the theoretical probability of landing on each section.

57) Describe the proportion of areas shown between the different shaded areas. For example, the largest un-shaded section is twice as large as the next largest section.

4 by 4 Square Dot Paper (p. 359)

58) Count the whole rectangle as 1 unit, then shade in given fractions.

59) Show all the ways that this 4 by 4 rectangle can be divided in half.

60) Choose one corner of the rectangle as the starting point of a path, and another corner as the end point. How many different ways are there to get from the start to the end without traveling twice through the same segment?

61) Students pair up, each with his or her own 4 by 4 dot paper. Students take turns drawing a figure without their partner seeing it and attempt to describe the figure to their partner, who then tries to recreate the figure. .

Half-inch graph paper (p. 360)

62) Folding the paper so that the grid shows on the outside, make boxes of 16 and 18 cubic inches in volume.

63) Use the fraction multiplication exercises in the fraction chapter to show various multiplications of fractions, such as $2\frac{1}{2} \times 1\frac{1}{2}$, etc.

64) Trace hands and other objects on top of the grid, then calculate the area and explain the steps taken to find the answers.

65) Create various common shapes by shading various squares (a house, a boat, etc.) Turn the designs into the teacher, who will read coordinate points based on these pictures. On a blank grid, using the left bottom part of the grid as the origin in the x-y axis (0,0), students draw lines based on the coordinates read by the teacher to recreate a student's design.

Quarter Inch Square Paper (p. 361)

66) To develop number sense, tape enough copies of the quarter-inch template on the wall to generate 10,000 and 100,000 small squares.

67) Determine how many sheets would be needed to show 1 million and 1 billion small squares.

69) Determine how much wall space would be needed to paste 1 million and 1 billion small squares.

70) Using a 10 by 12 rectangle within the grid, use different colors to show given percentages. Create a color legend along with the grid.

71) Using a 10 by 12 rectangle within the grid, use different colors to show given fractions. Create a color legend along with their grid.

Isometric Paper and Isometric Grids (p. 362-63))

72) Create various designs with sugar cubes, unifix cubes or other blocks. Using the isometric drawing samples in the geometry section, students draw the front, top and side views of the designs.

73) Using sugar cubes, unifix cubes or other type of blocks, create all possible buildings with a volume of 6 cubic units. The only restriction is that each cube must have one face perfectly aligned with the rest of the building. Afterwards students draw each building on the isometric paper.

74) Create all possible buildings with a surface area (not including the floor) of 13 and 17 square units. Again, the only restriction is that each cube must have one face perfectly aligned with the rest of the building.

Number Lines (p. 364)

75) Using a transparency of the number line template, label 1, 2 or 3 different whole numbers on the number line. Then label 3 or 4 entries with letters only. Discuss in groups the possible value of each letter from the information on the line.

76) Using a transparency of the number line template, label 1, 2 or 3 different decimal or fractional numbers on the number line. Then label 3 or 4 entries with letters only. Discuss in groups the possible value of each letter from the information on the line.

77) Using a transparency of the number line template, label 1, 2 or 3 different whole numbers on the number line. Then label 3 or 4 entries with letters only. Find the value of various algebraic expressions suing the letters (such as 2a-1, a +b-c, etc.).

Pictorial Worksheet (p. 365)

78) Using the examples from the addition of fractions exercises, create pictorial representations of various fraction addition computations (for example: $2\frac{1}{2} + 2\frac{1}{3}$).

79) Using the examples from the subtraction of fractions exercises, create pictorial representations of various fraction subtraction computations (for example: $3\frac{1}{4} - 1\frac{1}{3}$).

80) Using the examples from the multiplication of fractions exercises, create pictorial representations of various fraction multiplication computations (for example $3\frac{1}{2} \times 2\frac{1}{2}$).

Bibliography

Bebout, Harriett, C. *Using Children's Word-Problem Compositions for Problem-solving Instruction: A way to reach all children with mathematics.* In NCTM's Reaching All Students with Mathematics. Reston, VA. NCTM, 1993.

Bruner, J., S. *Toward A Theory Of Instruction.* Cambridge, MA. Harvard University Press, 1966.

Bruner, Jerome S. *The Process of Education.* Cambridge: Harvard University Press, 1960.

Burns, Marilyn. *About Teaching Mathematics.* Sausalito, CA. Math Solutions Publication, 2000.

California Department of Education. *Mathematics Framework for California Public Schools.* Sacramento, CA, 1999.

Chomsky, N. *Aspects Of The Theory Of Syntax.* Cambridge, MA. MIT Press, 1965.

Dewey, John. *Experience and Education.* New York: Macmillan Co., 1938.

Dienes, Zoltan P. *Building Up Mathematics.* Rev. ed. London: Hutchinson Educational, 1969.

English, D. Lyn. *Reasoning by Analogy: A Fundamental Process in Children's Mathematical Learning.* In NCTM's Developing Mathematical Reasoning in Grades K-12 (1999 Yearbook). Reston, Virginia, NCTM, 1999.

Glasersfeld, Ernst, von. *Learning as a Constructive Activity.* In Claude Janvier's (Editor) Problems of Representation In the Teaching and Learning of Mathematics. Hillsdale, New Jersey, LEA, 1987.

Herscovics, Nicolas. *Cognitive Obstacles Encountered in the Learning of Algebra.* In NCTM's Research Issues in the Learning and Teaching of Algebra. Reston, Virginia, NCTM, 1991

Janvier, Claude. *Translation Processes in Mathematics Education.* In Claude Janvier's (Editor) Problems of Representation In the Teaching and Learning of Mathematics. Hillsdale, New Jersey, LEA, 1987.

Kamii, Constance, and Washington, Mary, Ann. *Teaching Fraction: Fostering Children's Own Reasoning.* In NCTM's Developing Mathematical Reasoning in Grades K-12 (1999 Yearbook). Reston, Virginia, NCTM, 1999.

Kaput, James, J. *Linking Representations in the Symbol Systems of Algebra.* In NCTM's Research in the Learning and Teaching of Algebra. Reston, Virginia, NCTM, 1991.

Kaput, James, J. *Representation and Problem Solving: Methodological Issues Related to Modeling*. In Edward A. Silver's (Editor) Teaching and Learning Mathematical Problem Solving: Multiple Research Perspectives. Hillsdale, New Jersey, 1985.

Kaput, J. *Transforming algebra from an engine of inequity to an engine of mathematical power by "algebrafying" the K–12 curriculum*. In National Research Council (Ed.), *The nature and role of algebra in the K–14 curriculum: Proceedings of a national symposium*. Washington, DC: National Academy Press, 1998.

Kieran, Carolyn. *The Early Learning of Algebra: A Structural Perspective*. In NCTM's Research Issues in the Learning and Teaching of Algebra. Reston, Virginia, NCTM, 1991.

Kilpatrick, Jeremy. *A Retrospective Account of the Past 25 Years of Research on Teaching Mathematical Problem Solving*. In Edward A. Silver's (Editor) Teaching and Learning Mathematical Problem Solving: Multiple Research Perspectives. Hillsdale, New Jersey, 1985.

Lamon, Susan, J. *Teaching Fractions and Ratios For Understanding: Essential Content Knowledge and Instructional Strategies for Teachers*. Manwah, New Jersey. LEA, 1999.

Lesh, Richard, Post, Tom and Behr, Merlyn. *Representations and Translations Among Representations in Mathematics*. In Claude Janvier's (Editor) Problems of Representation In the Teaching and Learning of Mathematics. Hillsdale, New Jersey, LEA, 1987.

Lesh, Richard, Behr, Merlyn and Post, Tom. *Rational Number Relations and Proportions*. In Claude Janvier's (Editor) Problems of Representation In the Teaching and Learning of Mathematics. Hillsdale, New Jersey, LEA, 1987.

Lesh, Richard. *The Development of Representational Abilities in Middle School Mathematics*. In Sigels, Irving, E, (Editor) Development of Mental Representation: Theories and Applications. Mahwah, New Jersey. LEA, 1999.

Lesh, R., Hoover, M., Hole, B., Kelly, A., Post, T. *Principles for Developing Thought-Revealing Activities for Students and Teachers*. In A. Kelly, R. Lesh (Eds.), *Research Design in Mathematics and Science Education*. (pp. 591-646). Lawrence Erlbaum Associates, Mahwah, New Jersey, 2000.

Lesh, R., Cramer, K., Doerr, H., Post, T., Zawojewski, J. *Using a translation model for curriculum development and classroom instruction*. In Lesh, R., Doerr, H. (Eds.) *Beyond Constructivism. Models and Modeling Perspectives on Mathematics Problem Solving, Learning, and Teaching*. Lawrence Erlbaum Associates, Mahwah, New Jersey, 2003.

Mayer, Richard, E. *Implications of Cognitive Psychology for Instruction in Mathematical Problem Solving*. In Edward A. Silver's (Editor) Teaching and Learning Mathematical Problem Solving: Multiple Research Perspectives. Hillsdale, New Jersey, 1985.

National Council of Teachers of Mathematics. *Professional Standards For Teaching Mathematics*. Reston, VA, 1995.

Piaget, J. *The Psychology Of Intelligence*. London. Routledge & Kegan, 1950.

Piaget, J. *Play, Dreams And Imitation In Childhood*. New York. Norton, 1962.

Post, T. *The Role of Manipulative Materials in the Learning of Mathematical Concepts*. In Selected Issues in Mathematics Education. Berkeley, CA: National Society for the Study of Education and National Council of Teachers of Mathematics, McCutchan Publishing Corporation, 1981.

Reys, Robert E., and Post, Thomas R. *The Mathematics Laboratory: Theory to Practice*. Boston: Prindle, Weber, and Schmidt, 1973.

Reys, et al. *Helping Children Learn Mathematics*. New York, John Wiley & Sons, Inc., 2001.

Romberg, Thomas, A. and Kaput, James, J. *Mathematics Worth Teaching, Mathematics Worth Understanding*. In Fennema, Elizabeth, and Romberg, Thomas, A. (Editors) Mathematics Classrooms That Promote Understanding. Mahwah, New Jersey, LEA, 1999.

Stein, Mary Kay et Al. *Implementing Standards-Based Mathematics Instruction*. New York, Teachers College Press, 2000.

Usiskin, Zalman. *Conceptions of School Algebra and Uses of Variables*. In NCTM's The Ideas of Algebra, K-12 (1988 Yearbook). Reston, VA. NCTM, 1988.

Vygotsky, L., S. *Mind in society: The Development Of Higher Psychological Processes*. Cambridge, MA. Harvard University Press, 1978.

*Man's mind,
once stretched by a new idea,
never regains its original dimensions.*

~Oliver Wendell Holmes

Workshop and Training Information

Give your teachers what many are calling "the best professional development experience of their careers." Meaningful Learning is please to announce that you can book Guillermo Mendieta, author of Pictorial Mathematics, directly though us. Mr. Mendieta will guide your teachers on the most effective ways to use multiple representations to teach mathematic concepts meaningfully.

You may request professional development information via email at pictorialmath@yahoo.com or through our website, **www.pictorialmath.com**
.

You may also mail or fax the completed form below to get information on available professional development opportunities in your area.

Professional Development Mail Form

Meaningful Learning
P.O. Box 448
Etiwanda, Ca 91739
Tel: (909) 730-7312 – Fax (909) 854-5858

Name: _____ Date _____

School/District _____

Position: _____

Telephone _____ Email : _____

Address: _____

Please communicate your question/request in the space below:

_____

_____

_____

_____

Quick Order Form[1]

Make checks payable to **Meaningful Learning**

Order online at
www.PictorialMath.com
or use this form to order
by fax or mail

Mail or Fax to:
Meaningful Learning
P.O. Box 448
Etiwanda, CA 91739
(909) 730-7312 – Fax (909) 909-854-5858

Order online before
February 15th and
save even more!

We accept checks, credit cards and purchase orders (purchase orders must be faxed or mailed).

1-4 books: $34.95 per book
5-15 books: $31.45 per book (10% discount)
16-30 books: $29.70 per book (15% discount)
31+ books   Contact us

District orders of 100 books or more qualify for the free professional development offer. Call or email us for further details.

You can order online at PictorialMath.com

Name: _____

Institution: _____

Address: _____

City _____ State _____

Zip code: _____ Telephone: _____

Email: _____ Fax: _____

Number of Books	Price/Book (as detailed above)	Method of Payment	Subtotal
Taxes (add 7.75% if ordering within California - $2.70 for 1 book)			
Shipping* (3.50 for first book, $2.50 for every additional book)			
		Total	

www.PictorialMath.com

Live as if you were going to die tomorrow, learn as if you were to live forever.

Ghandi